Essential Mathematics for Life

BOOK 2

Decimals & Fractions

Fourth
Edition

GLENCOE
McGraw-Hill

New York, New York
Columbus, Ohio
Mission Hills, California
Peoria, Illinois

Authors

Mary S. Charuhas
Associate Dean
College of Lake County
Grayslake, Illinois

Dorothy McMurtry
District Director of ABE,GED,
   ESL
City Colleges of Chicago
Chicago, Illinois

The Mathematics Faculty
American Preparatory Institute
Killeen, Texas

Contributing Writers

Kathryn S. Harr
Mathematics Instructor
Pickerington, Ohio

Priscilla Ware
Educational Consultant and
   Instructor
Columbus, Ohio

Dr. Pearl Chase
Professional Consultants of Dallas
Cedar Hill, Texas

Contributing Editors and Reviewers

Barbara Warner
Monroe Community College
Rochester, New York

Michelle Heatherly
Coastal Carolina Community
   College
Jacksonville, North Carolina

Anita Armfield
York Technical College
Rock Hill, South Carolina

Judy D. Cole
Lafayette Regional Technical
   Institute
Lafayette, Louisiana

Mary Fincher
New Orleans Job Corps
New Orleans, Louisiana

Cheryl Gunderson
Rusk Community Learning
   Center
Ladysmith, Wisconsin

Cynthia A. Love
Columbus City Schools
Columbus, Ohio

Joyce Claar
South Westchester BOCES
Valhalla, New York

John Grabowski
St. Joseph Hill Academy
Staten Island, New York

Virginia Victor
Maple Run Youth Center
Cumberland, Maryland

Sandi Braga
College of South Idaho
Twin Falls, Idaho

Maggie Cunningham
Adult Education
Schertz, Texas

Sylvia Gilliard
Naval Consolidated Brig
Charleston, South Carolina

Eva Eaton-Smith
Cecil Community College
Elkton, Maryland

Fabienne West
John C. Calhoun State
   Community College
Decatur, Alabama

Photo credits: Cover, Ralph Mercer/Tony Stone Images; 12, Brent Turner/BLT Productions; 14, Roger K. Burnard; 17, 36, 46, Matt Meadows; 54, Doug Martin; 61, Brent Turner/BLT Productions; 70, Aaron Haupt; 84, Spencer Grant/Stock Boston; 103, Matt Meadows; 104, 132, file photo; 135, Matt Meadows; 147, S. Barth/H. Armstrong Roberts; 158, 159, Matt Meadows; 166, Brent Turner/BLT Productions.

Send all inquiries to:
Glencoe/McGraw-Hill
936 Eastwind Drive
Westerville, Ohio 43081

ISBN: 0-02-802609-8

1 2 3 4 5 6 7 8 9 POH 02 01 00 99 98 97 96 95 94

C O N T E N T S

Decimals

Unit 1 Review of Whole Numbers

Unit 2 Meaning of Decimals

Unit 3   Adding and Subtracting Decimals

Unit 4 Multiplying and Dividing Decimals

Unit 5   Meaning of Fractions

Unit 6   Mixed Numbers

Unit 7   Multiplying and Dividing Fractions and Mixed Numbers

Unit 8 Adding and Subtracting Fractions and Mixed Numbers

Unit 9 Changing Fractions to Decimals and Decimals to Fractions

Review of Whole Numbers

Find the value of each expression.

1. $(36 \div 4) + 5 = $ _____   **2.** $(3 \times 2) \times 7 = $ _____   **3.** $(6 + 8) - 9 = $ _____

4. $6 + (56 \div 7) = $ _____   **5.** $16 - (3 \times 3) = $ _____   **6.** $49 \div (3 + 4) = $ _____

Read each number. Write the place value of each underlined digit.

7. 8,6<u>2</u>7 _____   **8.** 25<u>7</u> _____

9. 642,6<u>1</u>6 _____   **10.** 84,<u>8</u>64 _____

11. <u>5</u>,465,854 _____   **12.** <u>5</u>08,264 _____

Compare the following, using >, <, or =.

13. 415 _____ 398   **14.** 6,038 _____ 6,039   **15.** 46,873 _____ 4,698

16. 367,934 _____ 369,734   **17.** 1,999 _____ 8,101   **18.** 567 _____ 576

19. 16,378 _____ 16,783   **20.** 353,960 _____ 36,897   **21.** 7,638 _____ 7,549

Round each number to the place named.

22. 5,674 (nearest ten)

_____

23. 75,546 (nearest thousand)

_____

24. 356,406 (nearest ten thousand)

_____

25. 326,975 (nearest hundred thousand)

_____

Write the whole numbers in order from largest to smallest.

26. 2,114; 2,410; 3,000; 2,069

_____

27. 1,536; 1,510; 1,551; 1,531

_____

Solve the following.

28.
```
  9 8
- 3 5
```

29.
```
  3 4
  6 1
+ 5 3
```

30.
```
  8 9
×  4
```

31. $8\overline{)49}$

32.
```
    4 3
    7 2
+ 1 3 7
```

33.
```
  7 0 9
×   1 1
```

34. $74\overline{)17,300}$

35.
```
  6 0 4
- 1 7 2
```

36.
```
  8 4 1
× 1 2 4
```

37. $434\overline{)153,754}$

38.
```
  6 0 1 , 1 0 5
  1 7 7 , 0 4 3
    2 0 , 3 4 1
+ 5 4 3 , 0 0 1
```

39.
```
  5 , 7 6 2 , 1 1 5
- 4 , 2 3 5 , 0 0 9
```

40.
```
  4 , 3 8 7 , 5 0 0
      2 7 , 5 4 1
      9 0 , 8 5 5
+   2 2 5 , 4 2 1
```

41.
```
  5 , 7 1 3
×   1 4 1
```

Use estimation to solve the following. Then, find the actual answers.

42.
$$\begin{array}{r} 7\,5\,8 \\ +\,2\,4\,3 \\ \hline \end{array}$$

43.
$$\begin{array}{r} 5\,,5\,3\,1 \\ -\,1\,,3\,7\,9 \\ \hline \end{array}$$

44.
$$\begin{array}{r} 7\,7\,7 \\ \times\,1\,0\,9 \\ \hline \end{array}$$

45. $68\overline{)1,440}$

Solve the following problems. Circle the correct answer.

46. A bolt of fabric is 63 yards long and 54 inches wide. Seven customers want equal amounts. How many yards will each get?

(1) 2   (2) 6

(3) 7   (4) 9

(5) 5

47. Team A sold 15 fewer tickets to the school play than Team B. Together the two teams sold 125 tickets. How many tickets did Team A sell?

(1) 15   (2) 55

(3) 70   (4) 110

(5) 85

48. A box of candy, which contains 36 candy bars, costs $5 per box. If Sofia purchases 612 candy bars, how much does she have to pay?

(1) $36   (2) $85

(3) $56   (4) $180

(5) $97

49. A produce truck has 373 crates of various products to deliver to three stores. One store ordered 173 crates, the second ordered 54. What was the order for the third store?

(1) 146   (2) 173

(3) 200   (4) 319

(5) 276

50. A crew of 17 firefighters discovered a forest fire at 3:30 p.m. on Tuesday. They finally brought the fire under control at 4:30 a.m. on Wednesday. If each crew member took a two-hour break, how many person hours were spent to control the fire after it was discovered?

(1) 13   (2) 77

(3) 187   (4) 221

(5) 154

51. Mi Ling is putting nine photos on each page of his family album. He has filled 37 pages. If he has 45 more photos to put in the album, how many pages of photos will be filled?

(1) 42   (2) 46

(3) 54   (4) 82

(5) 73

Order of Operations Using Basic Facts

Some expressions have more than one operation. The parentheses ( ) tell which operation to do first. After working inside the parentheses, you must do multiplication and division before addition and subtraction.

Examples

A. Add first.
$(4 + 5) - 3$
$\quad 9 \quad - 3 = 6$

B. Subtract first.
$9 + (5 - 3)$
$9 + \quad 2 \quad = 11$

C. Multiply first.
$(4 \times 5) \div 2$
$\quad 20 \quad \div 2 = 10$

D. Divide first.
$(9 \div 3) + 4$
$\quad 3 \quad + 4 = 7$

> **MATH HINT**
>
> **P**EMDAS will help you remember the correct order. P parentheses, E exponents, MD multiplication and division left to right, AS addition and subtraction left to right.

Practice

Do these operations in the order shown by the parentheses. Find the value of each expression.

1. $(15 \div 5) + 9 = $ _____

2. $(6 - 4) \times 6 = $ _____

3. $8 + (10 - 4) = $ _____

4. $12 - (54 \div 9) = $ _____

5. $8 + (6 \times 3) = $ _____

6. $(13 - 6) \times 5 = $ _____

7. $(63 \div 7) \times 9 = $ _____

8. $72 - (15 - 8) = $ _____

9. $(21 \div 3) \times 4 = $ _____

10. $(16 - 8) \times 5 = $ _____

11. $17 - (40 \div 8) = $ _____

12. $(45 \div 9) - 5 = $ _____

Place parentheses around the operation that was done first.

13. $8 \times 5 + 3 = 43$

14. $15 - 9 \times 9 = 54$

15. $4 \times 2 + 5 = 13$

16. $4 + 2 \times 3 = 10$

17. $3 + 5 \times 4 = 23$

18. $18 \div 6 + 3 = 2$

19. $4 + 2 \times 3 = 18$

20. $3 + 5 \times 4 = 32$

21. $18 \div 6 + 3 = 6$

Place Value, Ones, Tens, and Hundreds

The digits 0, 1, 2, 3, 4, 5, 6, 7, 8, and 9 are the only symbols needed for writing all whole numbers.

Examples

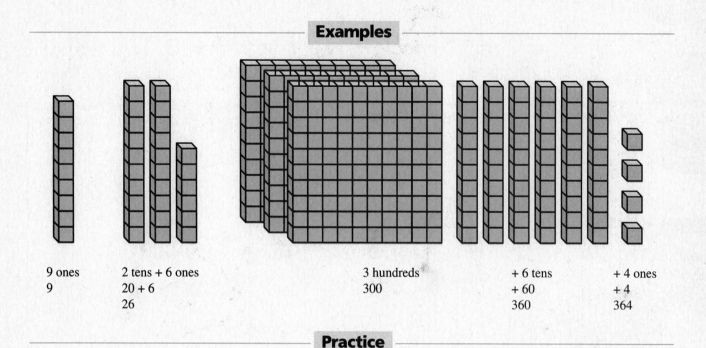

9 ones	2 tens + 6 ones	3 hundreds	+ 6 tens	+ 4 ones
9	20 + 6	300	+ 60	+ 4
	26		360	364

Practice

Read each number. Give the place value of each underlined digit.

1. 58<u>2</u> _____

2. <u>9</u>87 _____

3. 71<u>1</u> _____

4. <u>4</u>58 _____

5. 1<u>5</u> _____

6. 5<u>0</u>4 _____

7. <u>6</u>83 _____

8. 4<u>5</u> _____

Write the number in the list which has

9. 8 tens _____

10. 8 hundreds _____

11. 8 ones _____

12. 6 ones _____

> 9,856
> 8,749
> 7,698
> 6,987

Place Value, Thousands, Ten Thousands, Hundred Thousands, and Millions

The population of the United States is 248,709,873. Large numbers are sometimes difficult to read. The chart below will help you read large numbers.

millions			thousands			ones		
hundred millions	ten millions	millions	hundred thousands	ten thousands	thousands	hundreds	tens	ones
2	4	8	7	0	9	8	7	3

Practice

Read each number. Give the place value of each underlined digit.

1. 6<u>4</u>2,297

2. <u>9</u>0,768

3. <u>1</u>,406,741

4. <u>4</u>49,616

5. <u>5</u>08,264

6. 84,<u>8</u>64

7. 517,<u>8</u>86

8. 1,1<u>8</u>7,976

Write the number in the list which has

9. 6 tens _____

10. 3 thousands _____

11. 5 hundred thousands _____

12. 8 millions _____

> 632
> 5,170
> 53,704
> 1,521,251
> 8,245,064

Write the standard form for

13. fifty-four thousand, two hundred seventy-six _____

14. eighty-six thousand, three hundred four _____

15. twenty-seven thousand _____

16. three hundred sixty-seven thousand, nine hundred thirteen _____

Comparing and Ordering Whole Numbers

Here are the signs used when comparing numbers: $<$, $>$, and $=$.

7 > 5 means   5 < 7 means   (5 + 3) = 8 means

7 is **greater than** 5.   5 is **less than** 7.   (5 + 3) is **equal to** 8.

To **compare large numbers,** compare the digits that have the largest place value. If these digits are the same, compare digits in the next place and so on until one digit is greater. Compare from left to right.

Example

Which is greater, 356 or 365?

Look at the hundreds place.   Look at the tens place.

356   365   356   365

The 3's are the same.   5 < 6.   So, 356 < 365.

Practice

Compare the following, using >, <, or =.

1. 15 _____ 19

2. 297 _____ 295

3. 415 _____ 445

4. 3090 _____ 3081

5. 4328 _____ 4438

6. 8635 _____ 8321

7. 15,687 _____ 14,687

8. 25,617 _____ 24,569

9. 374,900 _____ 398,900

10. 290,351 _____ 272,111

11. 424,371 _____ 422,371

12. 732,000 _____ 723,511

13. 2,341,357 _____ 1,373,621

14. 5,251,631 _____ 6,251,000

Compare the following using >, <, or =.

15. $(12 \div 3) + 9$ _____ $(36 \div 4) + 5$

16. $11 + (11 - 4)$ _____ $(3 \times 2) + 7$

17. $20 - (40 \div 5)$ _____ $(12 - 8) \times 9$

18. $(9 \times 4) + (3 \times 2)$ _____ $(8 \times 6) - 6$

19. $(4 \times 5) \div 2$ _____ $(2 \times 8) \div 4$

20. $4 + (7 + 3)$ _____ $(6 \times 10) \div 5$

21. $(18 \div 3) \times 8$ _____ $7 \times (3 + 3)$

22. $28 + 5$ _____ $(3 \times 2) + 5$

23. $(36 \div 9) + 20$ _____ $(35 \div 5) + 20$

24. $(5 + 1) + 6$ _____ $(72 \div 8) - (4 - 1)$

25. $(5 \times 18) \div (3 \times 5)$ _____ $12 + (8 - 3)$

26. $(25 \div 5) + (6 + 3)$ _____ $25 - (2 \times 6)$

Write the whole numbers in order from largest to smallest.

27. 76, 86, 74, 69 _____

28. 119, 121, 106, 116 _____

29. 2,264; 2,335; 2,291; 2,912 _____

30. 498, 597, 488, 700 _____

31. 408,587; 488,700; 48,900 _____

32. 189,501; 110,999; 103,273 _____

Write the whole numbers in order from smallest to greatest.

33. 383, 317, 371, 357 _____

34. 8,888; 8,686; 8,863; 8,648 _____

35. 1,717; 1,770; 1,707; 1,771 _____

36. 512,675; 521,675; 512,876 _____

Rounding Whole Numbers

Odometers show distances traveled. To estimate distances, the mileage on the odometer can be rounded. To round to any place, look at the digit to the right of the place you want to round. Round up if the digit is 5 or greater than 5. Keep the place number if the digit is less than 5. Everything to the right of the place becomes zero.

Examples

The odometer of a car reads 123,384 miles. Round the distance to the nearest ten, hundred, thousand, ten thousand, and hundred thousand.

A. 123,384 rounded to the nearest ten is 123,380.
B. 123,384 rounded to the nearest hundred is 123,400.
C. 123,384 rounded to the nearest thousand is 123,000.
D. 123,384 rounded to the nearest ten thousand is 120,000.
E. 123,384 rounded to the nearest hundred thousand is 100,000.

Round these numbers to the nearest tens, hundreds, and thousands.

	tens	hundreds	thousands
1. 8,443	_____	_____	_____
2. 7,685	_____	_____	_____
3. 3,655	_____	_____	_____

Round these numbers to the nearest ten thousands and hundred thousands.

	ten thousands	hundred thousands
4. 754,386	_____	_____
5. 326,175	_____	_____
6. 166,999	_____	_____

Adding and Subtracting Whole Numbers

Addition:
When two or more numbers (**addends**) are added together, the
answer is called the **sum** or **total.**

Subtraction:
The **minuend** is the top number in the subtraction problem.
The **subtrahend** is the bottom number in the subtraction problem.
The **difference** is the answer to a subtraction problem.

Examples

A.
```
  7   Addend
+ 1   Addend
  8   Sum
```

B.
```
  4
  7
  8
+ 5
 24
```

C.
```
  79   Addend
+ 62   Addend
 141   Total
```

D.
```
  7,305
+ 3,387
 10,692
```

E.
```
  348   Minuend
- 135   Subtrahend
  213   Difference
```

F.
```
  974
- 482
  492
```

Practice

Add or subtract.

1.
```
  73
  54
  91
+ 40
```

2.
```
  97
  83
  84
+ 85
```

3.
```
  70
  21
  64
+ 75
```

4.
```
  93
  46
  79
+ 70
```

5.
```
  59
- 23
```

6.
```
  52
- 49
```

7.
```
  349
-  56
```

8.
```
  901
- 471
```

Add.

9. $\begin{array}{r} 76 \\ +97 \\ \hline \end{array}$

10. $\begin{array}{r} 45 \\ +43 \\ \hline \end{array}$

11. $\begin{array}{r} 79 \\ +62 \\ \hline \end{array}$

12. $\begin{array}{r} 34 \\ +60 \\ \hline \end{array}$

13. $\begin{array}{r} 965 \\ +249 \\ \hline \end{array}$

14. $\begin{array}{r} 563 \\ +878 \\ \hline \end{array}$

15. $\begin{array}{r} 971 \\ +544 \\ \hline \end{array}$

16. $\begin{array}{r} 674 \\ +987 \\ \hline \end{array}$

17. $\begin{array}{r} 3,437 \\ +6,727 \\ \hline \end{array}$

18. $\begin{array}{r} 5,263 \\ +9,876 \\ \hline \end{array}$

19. $\begin{array}{r} 3,387 \\ +5,798 \\ \hline \end{array}$

20. $\begin{array}{r} 8,459 \\ +3,595 \\ \hline \end{array}$

21. $\begin{array}{r} 6,876 \\ 4,049 \\ +11 \\ \hline \end{array}$

22. $\begin{array}{r} 3,695 \\ 3,746 \\ +1,880 \\ \hline \end{array}$

23. $\begin{array}{r} 9,876 \\ 7,254 \\ +957 \\ \hline \end{array}$

24. $\begin{array}{r} 5,935 \\ 7,627 \\ +177 \\ \hline \end{array}$

Subtract.

25. $\begin{array}{r} 939 \\ -368 \\ \hline \end{array}$

26. $\begin{array}{r} 695 \\ -265 \\ \hline \end{array}$

27. $\begin{array}{r} 348 \\ -182 \\ \hline \end{array}$

28. $\begin{array}{r} 673 \\ -198 \\ \hline \end{array}$

29. $\begin{array}{r} 7,365 \\ -487 \\ \hline \end{array}$

30. $\begin{array}{r} 9,725 \\ -981 \\ \hline \end{array}$

31. $\begin{array}{r} 5,851 \\ -674 \\ \hline \end{array}$

32. $\begin{array}{r} 7,366 \\ -569 \\ \hline \end{array}$

33. $\begin{array}{r} 9,000 \\ -7,068 \\ \hline \end{array}$

34. $\begin{array}{r} 4,000 \\ -2,170 \\ \hline \end{array}$

35. $\begin{array}{r} 7,800 \\ -5,113 \\ \hline \end{array}$

36. $\begin{array}{r} 5,200 \\ -5,167 \\ \hline \end{array}$

37. $\begin{array}{r} 93,947 \\ -56,284 \\ \hline \end{array}$

38. $\begin{array}{r} 67,245 \\ -37,849 \\ \hline \end{array}$

39. $\begin{array}{r} 4,000 \\ -1,485 \\ \hline \end{array}$

40. $\begin{array}{r} 3,675 \\ -2,941 \\ \hline \end{array}$

Problem Solving—Addition and Subtraction of Whole Numbers

Word problems are not difficult to master if you keep in mind the following steps:

Step 1   Read the problem and underline the key words. These words will generally relate to some mathematics reasoning computation.

Step 2   Make a plan to solve the problem. Ask yourself, Should I add, subtract, multiply, divide, round, or compare? You may have to do more than one of these operations for the same problem.

Step 3   Find the solution. Use your math knowledge to find your answer.

Step 4   Check the answer. Ask yourself, Is this answer reasonable? Did you find what you were asked for?

Some key words you may find are:

Addition	Subtraction	Multiplication	Division
sum	difference	product	quotient
total	remainder	times	how many
increase	how much more	of	average
together	how much less	apiece	shared
both	decreased by	multiplied by	
altogether	diminished by		

Examples

A. Ono had 300 telephone books to deliver. At the end of the day, he had delivered 217 books. How many books remain to be delivered?

Step 1   Determine how many books remain at the end of the day. The key word is **remain.**

Step 2   The key word **remain** or **remainder** is a subtraction word. To solve the problem, you will have to subtract the number of books delivered from the number of books at the beginning of the day.

Step 3 Find the solution.

    3 0 0 books at the beginning of the day
    − 2 1 7 books delivered
      8 3 books remain

Step 4 Check the answer. Does it make sense that Ono had 83 books remaining at the end of the day? Yes, the answer is reasonable.

B. Roberta is on a medically supervised weight control program. She must exercise for 35 minutes a day and consume no more than 1600 calories a day. Her breakfast meal totaled 365 calories. The sum of her calories for lunch was 595. How many calories remain for dinner?

Step 1 Determine how many calories Roberta can have for dinner. Three key words in this problem are **totaled, sum,** and **remain.**

Step 2 The key words indicate which operations should occur—addition and subtraction.

Step 3 Find the solution.

    3 6 5 calories for breakfast   1 6 0 0 daily calories allowed
    + 5 9 5 calories for lunch   −   9 6 0 calories consumed
      9 6 0 total calories   6 4 0 calories remain for dinner

Step 4 Check the answer. Does it make sense that Roberta's dinner should be no more than 640 calories? Yes, the answer is reasonable.

Practice

Problem Solving

Solve the following, using the steps for solving word problems.

1. The nursery has 28 pine trees, 39 maple trees, 41 oak trees, 54 birch trees, and 65 elm trees. How many trees does the nursery have in all?

              _____

2. The Quick Stop gas station sold 3,115 gallons of fuel one weekend. Of that amount, 1,223 gallons were premium unleaded and 156 gallons were diesel. If the remaining gallons of fuel were unleaded regular, how many gallons were sold of unleaded regular?

              _____

3. The seating capacity at the football stadium is 57,782. At the last game 42,987 people were recorded as paid and an additional 1,876 people had passes. How many seats were empty?

_____

4. Carrie has set a goal of saving $600 in her savings account. She has a balance of $450 in the account. If she deposits $53 today, how far away is she from her goal?

_____

5. The Town Players gave three performances. There were 921 people at the first performance and 539 at the second. If 1,987 people saw the play, how many people were at the third performance?

_____

6. Felipe and Maria are planning their wedding. They ordered 375 invitations. Felipe's list has 122 guests, and Maria's list has 239. How many invitations are unused?

_____

7. The shopping center has 2,050 parking spaces. On Monday there were 275 fewer cars than available spaces. How many cars were parked?

_____

8. In the local school board election, Candidate A received 3,435 votes. Candidate B received 8,342 votes. Candidate C received 1,500 more votes than Candidate B. How many votes were cast in the election?

_____

9. In December, a local hamburger chain sold 4,354 hamburgers. This was 1,567 fewer than in November. What was the total number of hamburgers sold for the two months?

_____

10. Village Hardware ordered 250 wrench sets. On Monday, the store received 100 sets. By Friday, another 45 sets had arrived. How many more sets are needed to complete the order?

_____

Multiplying and Dividing Whole Numbers

Multiplication:
The **multiplicand** is the number that is multiplied by another number.
The **multiplier** is the number that multiplies another number.
Sometimes the **multiplicand** and the **multiplier** are called **factors**.
The **product** is the answer to a multiplication problem.

Division:
The **dividend** is the number that is divided by another number.
The **divisor** is the number that divides another number.
The **quotient** is the answer to a division problem.
The **remainder** is the amount left over as undivided at the end of a division problem.

Examples

A.
```
    2 9   multiplicand
  ×   2   multiplier
    5 8   product
```

B.
```
      3 0 4   multiplicand
    ×   5 7   multiplier
    2 1 2 8
  1 5 2 0
  1 7 , 3 2 8   product
```

C.
```
              963   quotient
divisor   7)6,741   dividend
          −6 3
            44
           −42
            21
           −21
```

D.
```
               900R2   quotient
divisor   52)46,802    dividend
             −46 8
                 0
                −0
                 2
                −0
                 2
```

Multiply.

1. $\begin{array}{r} 34 \\ \times\ 13 \\ \hline \end{array}$

2. $\begin{array}{r} 53 \\ \times\ 72 \\ \hline \end{array}$

3. $\begin{array}{r} 64 \\ \times\ 26 \\ \hline \end{array}$

4. $\begin{array}{r} 26 \\ \times\ 58 \\ \hline \end{array}$

5. $\begin{array}{r} 708 \\ \times\ 69 \\ \hline \end{array}$

6. $\begin{array}{r} 505 \\ \times\ 35 \\ \hline \end{array}$

7. $\begin{array}{r} 765 \\ \times\ 50 \\ \hline \end{array}$

8. $\begin{array}{r} 507 \\ \times\ 93 \\ \hline \end{array}$

9. $\begin{array}{r} 494 \\ \times\ 27 \\ \hline \end{array}$

10. $\begin{array}{r} 796 \\ \times\ 11 \\ \hline \end{array}$

11. $\begin{array}{r} 900 \\ \times\ 13 \\ \hline \end{array}$

12. $\begin{array}{r} 905 \\ \times\ 39 \\ \hline \end{array}$

13. $\begin{array}{r} 1,845 \\ \times\ 36 \\ \hline \end{array}$

14. $\begin{array}{r} 5,639 \\ \times\ 24 \\ \hline \end{array}$

15. $\begin{array}{r} 2,573 \\ \times\ 29 \\ \hline \end{array}$

16. $\begin{array}{r} 4,007 \\ \times\ 14 \\ \hline \end{array}$

Divide.

17. $8\overline{)746}$

18. $8\overline{)6,584}$

19. $9\overline{)7,677}$

20. $7\overline{)3,423}$

21. $23\overline{)1,794}$

22. $34\overline{)2,959}$

23. $45\overline{)4,005}$

24. $56\overline{)4,256}$

25. $60\overline{)2,280}$

26. $28\overline{)4,032}$

27. $23\overline{)3,085}$

28. $24\overline{)4,944}$

Use the steps outlined on page 12 to solve the following problems.

29. The owners of a fruit stand sold 128 bushels of apples at $6 per bushel, 42 bushels of pears at $8 each, and 36 bushels of peaches for $432. What was the price per bushel for the peaches?

_____

30. It is estimated that a family-run variety store profits $2,846 a month. What is the annual profit of the store if it is closed one month per year?

_____

31. The Larkspur Cove Day Care Center bought 14 blankets at $5 each, 3 rocking horses for $180, and 6 painting easels for $17 each. When the bill arrived, the center director chose to pay the bill in four equal payments. How much will each payment be?

_____

32. Five brothers want to buy a video game and game cartridge. Three brothers have $45 each to spend. The two remaining brothers decide to split the balance equally. If the total cost of the video game and cartridges is $255, how much will the two brothers each need to contribute?

_____

33. Cara has collected 160 pairs of mittens and an equal number of scarves and hats for children at four homeless shelters. If divided equally, how many items will each shelter receive?

_____

34. One regional airline has a passenger plane with three seating sections—first class, business, and coach. There are 10 seats in first class and 24 seats in business. If the plane's passenger capacity is 330, how many seats are there in coach?

_____

35. The computer learning center at the local community college has 8 terminals that seat 6 people, 24 terminals that seat 4 people, and 20 terminals that seat 2 people. How many people can be in the lab at once?

_____

Posttest

Find the value of each expression.

1. $(63 \div 7) + 2 =$ _____   2. $(4 \times 3) \times 6 =$ _____   3. $(5 + 7) - 9 =$ _____

4. $12 + (14 \div 7) =$ _____   5. $24 - (2 \times 9) =$ _____   6. $48 - (8 + 4) =$ _____

Read each number. Write the place value of each underlined digit.

7. 7,2<u>8</u>6 _____   8. 58<u>0</u> _____

9. 975,<u>9</u>49 _____   10. 51,<u>5</u>31 _____

11. 1,<u>6</u>87,076 _____   12. <u>4</u>98,153 _____

Compare the following, using >, <, or =.

13. 405 _____ 449   14. 360,499 _____ 361,000   15. 46,873 _____ 4,698

16. 3,090 _____ 2,999   17. 8,701 _____ 7,802   18. 5,196 _____ 5,164

19. 365,000 _____ 356,000   20. 54,101 _____ 54,079   21. 9,999 _____ 10,001

Round each number to the place named.

22. 7,401 (nearest ten)   23. 16,999 (nearest thousand)

_____   _____

24. 5,655 (nearest thousand)   25. 326,975 (nearest ten thousand)

_____   _____

26. 754,386 (nearest hundred thousand)   27. 54,498 (nearest hundred)

_____   _____

Write the whole numbers in order from largest to smallest.

28. 9,475; 8,990; 9,587; 9,209

_____

29. 5,976; 5,967; 5,971; 5,987

_____

30. 29,035; 28,799; 29,110; 28,990

_____

31. 374,900; 385,800; 379,900

_____

Solve the following.

32. $\begin{array}{r} 5,739 \\ +\,1,806 \\ \hline \end{array}$

33. $\begin{array}{r} 50,548 \\ -\,31,462 \\ \hline \end{array}$

34. $\begin{array}{r} 607 \\ \times\,143 \\ \hline \end{array}$

35. $62\overline{)190}$

36. $42\overline{)9,743}$

37. $\begin{array}{r} 41 \\ 10,569 \\ +\,8,983 \\ \hline \end{array}$

38. $\begin{array}{r} 555,367 \\ -\,428,417 \\ \hline \end{array}$

39. $\begin{array}{r} 3,301 \\ \times\,279 \\ \hline \end{array}$

40. $\begin{array}{r} 5,233 \\ \times\,407 \\ \hline \end{array}$

41. $210\overline{)72,895}$

42. $\begin{array}{r} 371,407 \\ 450,892 \\ +\,119,738 \\ \hline \end{array}$

43. $\begin{array}{r} 4,931 \\ -\,1,479 \\ \hline \end{array}$

44. $286\overline{)637,290}$

45. $\begin{array}{r} 1,345,921 \\ 334,889 \\ 157,421 \\ +\,\quad 1,578 \\ \hline \end{array}$

Use estimation to solve the following. Then find the actual answers.

46. $\begin{array}{r} 7,043 \\ +\,1,993 \\ \hline \end{array}$

47. $\begin{array}{r} 5,763 \\ -\,2,941 \\ \hline \end{array}$

48. $\begin{array}{r} 3,259 \\ \times\,88 \\ \hline \end{array}$

49. $226\overline{)53,336}$

Read each problem. Circle the correct answer.

50. After a sale at Appleton's Speciality Shop, one inventory item included packages of T-shirts. A new order of 50 packages at $3 per shirt was received. If there are 12 T-shirts to a package, what was the new inventory of individual T-shirts?

(1) 150   **(2)** 196

(3) 336   **(4)** 600

(5) 36

51. One diet diary showed that a patient had eaten 12,348 calories in one week. The patient's doctor has prescribed a daily intake of 1,800 calories. What was the average daily calorie intake?

(1) 812   **(2)** 900

(3) 1,234   **(4)** 1,764

(5) 1,548

52. Hyde Parke's Food Cooperative orders 84 gallons of milk and 50 cartons of eggs for its members three times a week. At this rate, how many gallons of milk will be consumed in one year?

(1) 4,200   **(2)** 12,104

(3) 12,600   **(4)** 13,104

(5) 134

53. Carlos and Enrique share an apartment. Each pays an equal share for expenses. If Carlos earns $8 an hour and works 40 hours per week, how much money does he have left after paying his share of $1350 expenses per month?

(1) $70   **(2)** $160

(3) $605   **(4)** $675

(5) $1342

54. The local Orlando Literacy Project has 75 volunteers. Of these, 11 people have volunteered for 1–2 years, 15 have volunteered for 2–5 years, and 5 for more than 5 years. How many people have volunteered for less than a year?

(1) 42   **(2)** 44

(3) 49   **(4)** 64

(5) 51

55. A mid-sized car gets 18 miles per gallon of gas in city traffic and 23 miles to the gallon on the highway. Last month, a driver drove 1,564 miles on the highway and 414 miles in the city. How many gallons of gas were used during the month?

(1) 88   **(2)** 41

(3) 68   **(4)** 91

(5) 23

56. The Lyons Park District had 153 participants registered in all its spring session fitness exercise classes. Since the start of the session, 7 participants quit the morning class and 4 quit the afternoon class. If there are 139 participants left in all the classes, how many quit the evening class?

(1) 2   **(2)** 4

(3) 5   **(4)** 7

(5) 3

57. Kevin's annual fuel bill was $720. His electric and telephone bills averaged $50 per month. What was his average monthly expense for all three bills?

(1) $50   **(2)** $110

(3) $115   **(4)** $130

(5) $120

Write the place value of <u>2</u> in the following.

1. 38.492 _____

2. 38.249 _____

Write the decimal form for the following.

3. nine and five tenths _____

4. six dollars and eighty-eight cents _____

5. eighty-one thousandths _____

Circle the larger number in each pair.

6. 0.8 .08

7. 127.56 12.756

8. .05 .085

9. 0.0009 .00010

Write the decimals in order from the largest to the smallest.

10. 1.06 .106 10.6 _____ _____ _____

11. 8.4 8.04 .804 _____ _____ _____

Money and Decimals

Everyone who knows how to write dollars and cents knows how to write decimals. The dot in $5.50 is a **decimal point.** The dollars are to the **left** of the decimal point. The cents are to the **right** of the decimal point. The first digit to the right of the decimal point is the dime's position. The second digit to the right of the decimal point is the penny's position.

100¢/$1.00   50¢/$0.50   25¢/$0.25   10¢/$0.10   5¢/$0.05   1¢/$0.01

Examples

A. Write 2 dollars and 35 cents using the dollar sign and decimal point.

2 dollars and 35 cents = $2.35

B. Write 8 cents using the dollar sign and decimal point.

8 cents = $0.08

C. To pay $1.13, you need

1 dollar, 1 dime, 3 pennies.

D. To pay $14.09, you need

14 dollars, 0 dimes, 9 pennies.

Practice

Write the following, using the dollar sign and decimal point.

1. six dollars and ninety cents

_____

2. fifteen dollars and three cents

_____

3. nine cents

_____

4. ten cents

_____

How much is the 7 worth in each amount below?

5. $913.97   7 cents
_____

6. $7.89   _____

7. $710.20   _____

8. $156.70   _____

22

LIFE SKILL

Identifying Sums of Money

Write each sum of money using the dollar sign and decimal point.

1. $56.40

2. _____

3. _____

4. _____

5. _____

6. _____

Decimal Place Values

With dollars and cents, there are two decimal places to the right of the decimal point. When there is no dollar sign, the first place to the right of the decimal point is called **tenths**. The second place to the right of the decimal point is called **hundredths**.

Examples

When reading decimals **without** a whole number, use the following steps:

A. Read .7

Step 1   Read the number.   7
Step 2   Add the correct "*th*" word.   tenths

B. Read .47

Step 1   Read the number.   47
Step 2   Add the correct "*th*" word.
        hundredths

When reading a decimal **with** a whole number use the following steps:

C. Read 73.12

Step 1   Read the whole number.   73
Step 2   Read the decimal point as "and."   and
Step 3   Read the number to the right of the decimal point.   12
Step 4   Add the correct "th" word.   hundredths

Practice

Do these exercises orally. Ask your teacher to listen as you read.
Read the following decimals.

1. .2	**2.** 0.83	**3.** .02	**4.** 41.1
5. .09	**6.** 7.01	**7.** 527.3	**8.** 5964.15
9. 385.74	**10.** 11.01	**11.** 1.5	**12.** 518.3

In the place value chart below, four more places have been added to the right of the decimal point.

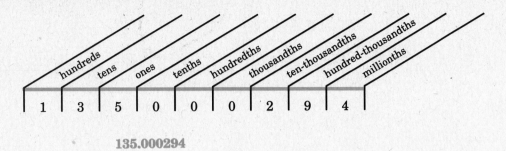

135.000294

one hundred thirty-five and two hundred ninety-four millionths

Practice

Finish these sentences.

1. The third place to the right of the decimal point is called _____

2. The fourth place to the right of the decimal point is called _____

3. The fifth place to the right of the decimal point is called _____

4. The sixth place to the right of the decimal point is called _____

Write as decimals.

5. fifty-seven and four tenths _____57.4_____

6. 312 ten-thousandths _____

7. seventy hundredths _____

8. 5,500 and five hundredths _____

9. 53 and fifty-three millionths _____

10. seven and seven hundred-thousandths _____

Write each of the numbers in words.

11. 15.18 _____

12. 13.00013 _____

13. .015 _____

14. 5.000097 _____

15. .7 _____

16. .0002 _____

Read each decimal in Column A. Match the place value of each underlined digit with the word(s) in Column B. The first one is done for you.

Column A

17. 4.451

18. 0.170516

19. 1,118.342

20. 4.1075

21. 806.215

22. 500.8342

23. 9,4270.6

24. 67.05761

25. 1,913.6785

26. 302.046507

Column B

a. hundredths

b. hundred-thousandths

c. millionths

d. tens

e. hundreds

f. ones

g. tenths

h. thousandths

i. ten-thousandths

j. thousands

Read each decimal below. Write the correct "th" word in the blank.

27. 200.51 _____

28. 57.6 _____

29. 2.854 _____

30. 175.6255 _____

31. 31,765.1 _____

32. 1.2875 _____

33. 7.138 _____

34. 175.63551 _____

35. 0.007654 _____

36. 2.88 _____

Locating Decimals on the Number Line

The points on the number line can be represented by whole numbers and decimals.

A.

The points on the Example A number line are represented by **whole numbers,** 2–7.

B.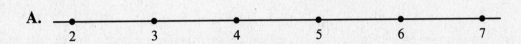

The whole numbers in Example B above are 5 and 6. The points between are represented by **tenths.**

C.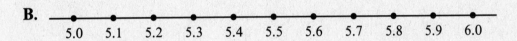

The beginning and ending points in Example C are 0.5 and 0.6. The points on Example C number line are represented in **hundredths.**

D.

1.110 1.111 1.112 1.113 1.114 1.115 1.116 1.117 1.118 1.119 1.120

The beginning and ending points in Example D are 1.110 and 1.120. The points between are represented in **thousandths.**

Practice

1. The whole numbers 4 and 8 are shown on the number line below. Write the whole numbers for each point indicated on the number line.

4   8

2. The whole numbers 6 and 7 are shown on the number line below. The points between 6 and 7 can be separated into tenths. Write the decimal for each point indicated on the number line.

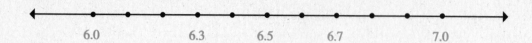

6.0   6.3   6.5   6.7   7.0

3. The beginning and ending points on the number line below are 0.7 and 0.8. The points between them are separated into hundredths. Write the decimal for each point indicated on the number line.

0.70   0.71   0.74   0.75   0.77   0.80

4. The beginning and ending points on the number line below are the whole numbers 2 and 3. The points between them are separated into thousandths. Write the decimal for each point indicated on the number line.

2.0   2.100   2.500   2.700   3.0

For problems 5 through 20, use the number lines to estimate the location of each point. The first two are done for you.

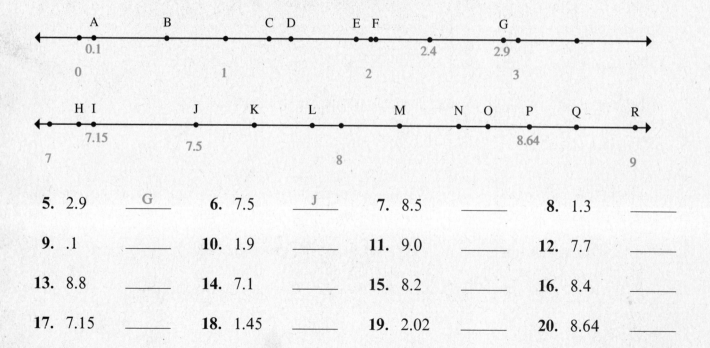

5. 2.9 _____G_____   **6.** 7.5 _____J_____   **7.** 8.5 _____   **8.** 1.3 _____

9. .1 _____   **10.** 1.9 _____   **11.** 9.0 _____   **12.** 7.7 _____

13. 8.8 _____   **14.** 7.1 _____   **15.** 8.2 _____   **16.** 8.4 _____

17. 7.15 _____   **18.** 1.45 _____   **19.** 2.02 _____   **20.** 8.64 _____

Comparing Decimals

Adding on or taking away zeros to the right of a decimal point does not change its value. The numbers are still equal.

The symbol **=** means "**is equal to.**"

---------- **Examples** ----------

These are equal decimals.

.4 = .40
4 tenths =
40 hundredths

A. 9.100 = 9.10 = 9.1 = 9.1000

B. 22.40 = 22.400 = 22.4 = 22.40000

If two decimals are not equal, one is greater than or less than the other.

The symbol **>** means "**is greater than.**"

The symbol **<** means "**is less than.**"

C. Compare .3 and .5.
   Use > or <.
   3 tenths is less than 5 tenths.
   .3 < .5

D. Compare .38 and .2.
   Use > or <.
   Add a zero to the right of .2.
   38 hundredths is greater than 20 hundredths.
   .38 > .2

---------- **Practice** ----------

Compare the following. Use >, <, or =.
The first one is done for you.

1. .3 ___>___ .19

2. 8 _____ .409

3. 9.2 _____ .092

4. .5 _____ .28

5. .3 _____ .0333

6. 1.7 _____ 1.669

7. 6.09 _____ .069

8. .8 _____ 1.0

9. .679 _____ .6790

10. .035 _____ .52

11. 4.87 _____ 48.7

12. .009 _____ .005

13. .0035 _____ .009

14. 17.1 _____ 1.717

29

15. 3.004 _____ 3.0040

16. .9 _____ .78

17. 7.3 _____ 73.00

18. .87 _____ .9

19. 5.6 _____ 5.8

20. 1.236 _____ 1.331

21. .0004 _____ .0040

22. 4.1 _____ 14

23. 0.0009 _____ 0.200

24. 0.049 _____ 0.490

25. 0.6 _____ 0.599

26. 39.6 _____ 39.60

27. .09 _____ 0.59

28. 1.586 _____ 01.5860

29. 15.9 _____ 159

30. 260.73 _____ 27.73

31. 8.41 _____ 08.41

32. .0214 _____ 0.0217

33. 4.02 _____ 3.96

34. 8 _____ 8.79

Which of the following statements is true? Circle the letter of your choice.

35. a. 3.12 is less than 3.02
    b. .006 is less than .06
    c. .652 is greater than .658
    d. 7.089 is greater than 7.09
    e. 5.75 is equal to 5.075

36. a. .12 is less than .02
    b. .0001 is less than .0001
    c. .256 is greater than 2.6
    d. 118.089 is greater than 118.09
    e. 0.560 is equal to .56

Which of the following statements is not true? Circle the letter of your choice.

37. a. .752 is greater than .0752
    b. .0007 is greater than .007
    c. 1.652 is greater than 1.05
    d. 1.652 is equal to 01.6520
    e. .0753 is less than 07.35

38. a. 3 dimes are .30 of a dollar
    b. 2 quarters are .50 of a dollar
    c. 4 nickels are .20 of a dollar
    d. 7 pennies are .07 of a dollar
    e. 3 nickels are .25 of a dollar

Writing Decimals in Order

When we write decimals in order from the largest to the smallest or the smallest to the largest, we must compare the place values.

─────────────────────── **Example** ───────────────────────

To write decimals in order from the smallest to the largest, follow these steps.

Step 1	Write the numbers in a column and line up the decimal points.	.1 2 .0 1 3 .1 1 5 .1 0 7		Step 2	Compare the tenths.	.1 2 .0 1 3 0 is smallest .1 1 5 .1 0 7

Step 3	When the tenths are the same, compare the hundredths.	.1 2 2 is
		    largest
		.1 1 5
		.1 0 7 0 is
		    smallest
				Step 4	Now arrange and complete the numbers.	.0 1 3
						.1 0 7
						.1 1 5
						.1 2 0

─────────────────────── **Practice** ───────────────────────

Write the decimals in order from smallest to largest.

1. .15   11.015   .2   .02   _____ _____ _____ _____

2. .23   .023   .32   .032   _____ _____ _____ _____

Write the decimals in order from largest to smallest.

3. .375   .0375   3.75   .00375   _____ _____ _____ _____

4. .125   .1025   .025   .0125   _____ _____ _____ _____

Writing Checks

When you write a check, use the following steps:

Step 1 Write the correct date in the upper right-hand corner.

Step 2 Write the name of the person or company you are paying.

Step 3 Write the money amount in numbers.

Step 4 Write the dollar amount in words and the cents as a fraction over 100.

Step 5 Sign your name.

Fill in the checks using the following data. The first one is done for you.

A. Mutual Life Insurance, $26.13

B. Forest Hills Apartments, $353.70

C. Commonwealth Edison, $56.39

D. National Telephone Co., $44.17

A.

Step 1 → April 14 19 95   7-15/520

Step 2 ↓   Step 3 →

PAY TO THE ORDER OF _Mutual Life Insurance_ $ 26.13

Twenty-six and ¹³/₁₀₀ ———— DOLLARS

City National Bank
Annapolis   Step 4

Memo _Policy #632_   _Betty Adair_

⑆052000159⑆   6343019⑈   Step 5

B.

_____ 19 ___   7-15/520

PAY TO THE ORDER OF _____ $ _____

_____ DOLLARS

City National Bank
Annapolis

Memo _____

⑆052000159⑆   6343019⑈

C.

_____ 19 ___   7-15/520

PAY TO THE ORDER OF _____ $ _____

_____ DOLLARS

City National Bank
Annapolis

Memo _____

⑆052000159⑆   6343019⑈

D.

_____ 19 ___   7-15/520

PAY TO THE ORDER OF _____ $ _____

_____ DOLLARS

City National Bank
Annapolis

Memo _____

⑆052000159⑆   6343019⑈

E. First Mortgage, $409.17

F. Home Shopper, $39.95

G. Fitness First Health Club, $24.95

H. Wilson Community College, $144.50

E.

_____ 19 _____ 7-15/520

PAY TO THE
ORDER OF _____ $ _____

_____ DOLLARS

City National Bank
Annapolis

Memo _____

⑆052000159⑆ 6343019⑈

F.

_____ 19 _____ 7-15/520

PAY TO THE
ORDER OF _____ $ _____

_____ DOLLARS

City National Bank
Annapolis

Memo _____

⑆052000159⑆ 6343019⑈

G.

_____ 19 _____ 7-15/520

PAY TO THE
ORDER OF _____ $ _____

_____ DOLLARS

City National Bank
Annapolis

Memo _____

⑆052000159⑆ 6343019⑈

H.

_____ 19 _____ 7-15/520

PAY TO THE
ORDER OF _____ $ _____

_____ DOLLARS

City National Bank
Annapolis

Memo _____

⑆052000159⑆ 6343019⑈

Rounding Decimals

Rounding decimals is similar to rounding whole numbers.

Example

To round a decimal to the nearest hundredth, follow these steps:

		55.717
Step 1	Underline the place digit to be rounded.	5 5 . 7 <u>1</u> 7
Step 2	Look at the digit to the right of the underlined digit. That digit is 7.	5 5 . 7 <u>1</u> 7
Step 3	If the digit to the right of the underlined digit is 5 or greater than 5, increase the underlined digit by 1. If the digit to the right of the underlined digit is less than 5, the underlined digit remains the same. The 7 is greater than 5, so the underlined digit is increased to 2.	
Step 4	All digits to the right of the underlined digit are dropped. The 7 in the thousandths position will be dropped.	5 5 . 7 <u>2</u> 7̶
Step 5	Write the rounded decimal.	5 5 . 7 2

Practice

Round each decimal to the nearest tenth. Then, round each decimal to the nearest hundredth.

Decimal to be rounded	nearest tenth	nearest hundredth
1. 0.0521	*0.1*	*0.05*
2. 17.5248		
3. 785.428		
4. 79.3672		
5. .739		
6. 66.175		
7. 99.8810		

Round to the nearest whole number.

8. 136.521 _____   9. 136.0251 _____   10. 6.58 _____

11. 6.258 _____   12. 18.36 _____   13. 18.63 _____

Round to the nearest tenth.

14. 0.38 _____   15. 0.34 _____   16. 519.38 _____

17. 43,500.19 _____   18. 0.71 _____   19. 0.74 _____

Round to the nearest hundredth.

20. 276.962 _____   21. 276.964 _____   22. 68.107 _____

23. 15.826 _____   24. 45.1291 _____   25. 86.584 _____

Round to the nearest thousandths.

26. 0.10768 _____   27. 2.2131 _____   28. .09476 _____

Round to the nearest ten-thousandths.

29. 58.23603 _____   30. 0.022941 _____   31. 13.90909 _____

Round to the nearest hundred-thousandths.

32. 10.875329 _____   33. 0.500573 _____   34. 429.304769 _____

Estimating the Cost of Groceries

When grocery shopping, you can use rounding of decimals to estimate the total cost of the items you are buying. An item marked $.59 can be rounded to $.60. An item that is marked $1.98 becomes $2.00. By rounding you can add faster as you shop.

There are two columns on the right. Column A has the actual price. In column B you are to give the rounded price. Round the decimals to either the nearest tenth or dollar amount. After giving the total estimated answer, give the exact dollars and cents amount. What is the difference between the two columns?

Column A Actual Price	Column B Rounded Price
	1. $2.00
$1.98	
	2. .50
.45	
	3. _____
.33	
	4. _____
1.79	
	5. _____
.71	
	6. _____
.69	
	7. _____
.62	
	8. _____
1.49	
	9. _____
1.13	
	10. _____
.81	
	11. _____
.51	
	12. _____
.52	
Total _____	Total _____

Difference between totals _____

Problem Solving—Comparing Decimals

Throughout this book and the other books in this series, you will be asked to solve word problems. Remember they are not difficult to master if you keep in mind the following steps:

Step 1   Read the problem and underline the key words. These words will generally relate to some mathematics reasoning computation.

Step 2   Make a plan to solve the problem. Ask yourself, Should I add, subtract, multiply, divide, round, compare? You may have to do more than one of these operations for the same problem.

Step 3   Find the solution. Use your math knowledge to find your answer.

Step 4   Check the answer. Ask yourself, Is the answer reasonable? Did you find what you were asked for?

Also, be alert for extra facts you do not need to solve the problem.

Example

Anastasia saw the perfect dress in a store window. She wants to buy it for a party she is going to next Saturday. The dress costs $35. She has $29 saved. Does she have enough money to buy the dress?

Follow these steps to solve this problem:

Step 1   Determine if Anastasia has enough money to buy the dress. The key words are: **costs** $35 and **saved** $29.

Step 2   This problem requires you to **compare** the cost amount with the amount saved. Is the saved amount of $29, greater than (>), less than (<), or equal to (=) the cost amount of $35?

Step 3   Find the solution. Which number is bigger? $35 > $29. Anastasia cannot afford to buy the dress.

Step 4   Check the answer. Does it make sense that $35 is greater than $29? Yes. Comparing these numbers on the number line would show that 35 is greater than 29.

37

Problem Solving

Solve the following, using the steps for solving word problems.

1. Ms. Jones has 45 students in her math class this year. Ms. Kasey has 38 students. Which teacher has fewer students this year?

    _____

2. A mutual fund reported a dividend of 5.954 shares to Mr. Estes. To the nearest whole number, how many shares of stock did he receive?

    _____

3. Brenda took five math tests. Her average came to 88.45. If Brenda's teacher rounds her grade to the nearest whole number, what is Brenda's math grade?

    _____

4. Jeff borrowed $5000.65 from the bank for a car. Henry borrowed $5000.58 to buy his car. And Denise borrowed $5001.65. Who borrowed the greatest amount? Who borrowed the least amount?

    _____

    _____

5. The Nogales family is saving to go on a family vacation. They have saved $400.63. They have determined that the trip will cost $400.45. Have they saved enough for the vacation trip?

    _____

Posttest

Write the following amounts using the dollar sign and decimal point.

1. 4 dollars and 59 cents _____

2. two cents _____

3. seventy cents _____

4. one dollar and five cents _____

5. 3 dollars and 3 cents _____

Name the place the 2 occupies in the decimals below.

6. 4.362 _____

7. 4.256 _____

8. 2.85 _____

9. .02673 _____

10. .00572 _____

Circle the decimals that are equal.

11. .02   .020   .20

12. 0.19   .109   .190

13. .3   .3000   .03

14. 6.66   .666   6.660

Use >, <, or = to compare the following decimals.

15. .256 _____ .3

16. .070 _____ .007

17. 52.1 _____ 521

18. 34.5 _____ 37.056

Write the decimals in order from the largest to the smallest.

19. 8.32   7.3   8.032

   _____   _____   _____

20. .05   .005   .0005

   _____   _____   _____

21. .0025   .505   .0505

   _____   _____   _____

Round to the place named.

22. 1,141.67 (tenths) _____

23. 1,141.5468 (thousandths) _____

24. 1,141.51781 (ten-thousandths) _____

Adding and Subtracting Decimals

Find the sums.

1. 5.464 + .01 + 129.2 _____

2. .0032 + 11.301 + 36 _____

3. 15.31 + .00151 + 8.18 _____

4. 5 + 13.615 + 4.01 _____

5.
```
  1 1 1 . 9 6 3
+   5 3 . 0 0 1 7
```

6.
```
  2 3 . 0 0 5 1
+   1 . 0 0 0 0 8
```

7.
```
  $   4 . 1 6
+   3 3 . 8 9
```

Find the differences.

8. 1 − .019 _____

9. $5.90 − $1.11 _____

10. .04 − .0178 _____

11.
```
   . 8
− . 4 0 6
```

12.
```
  $ 1 2 . 1 2
−     4 . 7 7
```

13.
```
  1 0
−   5 . 8 7 1
```

Solve the following problems.

14. In the spring, the park crew cut the rose bushes back to a height of 1.75 feet. At the end of summer, the bushes were measured at 3.7 feet. How many feet did the bushes grow?

_____

15. At the TriCity Triathlon, the athletes swim 2.5 miles, bike 27.35 miles, and run 10.07 miles. What is the total distance the athletes travel?

_____

Adding Decimals

Adding decimals is like adding whole numbers. To add decimals, follow these steps:

Step 1 Write the numbers in a column, lining up the decimal points. Put a decimal point on the right of any whole number.

Step 2 Add as with whole numbers.

Step 3 Bring the decimal point straight down into the answer.

Example

Add. 9.2 + .85 + .003 + 126.7 + 10 = ?

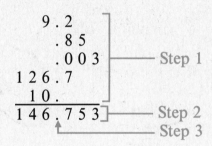

Practice

Add.

1.   1 . 2
       + 1 . 1

2.   1 . 5
        + 2 1 . 2

3.   1 2 6 . 5 4
             9 . 0 2 7
       +   . 0 9

4.   8 . 4 8
                . 0 0 3
           + 4 0 6 . 3

5.   3 3 . 5
              . 9 5
        + 7 1 3 . 7 9

6.   9 . 2 1
         2 7 5 . 3 2
       +   9 . 0 3 5

7. 11.73 + 82.737 + .0573 + .211

_____

8. 24.465 + 127.77   _____

9. 1.056 + 0.07875   _____

10. 166.73 + .373 + 5   _____

11. 3789 + 13.58 + 7.273   _____

Solve the following problems.

12. Maria and Stephen Valdez own a small clothing store. At the end of the day they had charge slips of $86.50, $43.20, $19.95, $12.09, and $17.59. What was the total amount of charges for that day?

_____

13. Mrs. Varca sent her son to the store for three items. They cost $.85, $1.49, and $1.19. How much did he spend?

_____

14. Lance works for the electric company. He fills the service trucks with gas. He pumped 15.7 gallons, 7.5 gallons, and 10.2 gallons. How many gallons did he pump in all?

_____

15. Mrs. Hutzelman has bills amounting to $23.27, $16.59, and $29.38. How much does she need in her checking account to pay all three?

_____

16. Kwan earns $197.25 a week as a word processor. She received a raise of $15.75 a week. What is her new salary?

_____

17. Arturo works as a waiter at the Wagon Wheel Restaurant. He keeps a record of his tips. How much did he earn in tips for 4 weeks? Use the lists below to help you answer the question.

_____

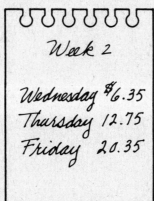

Week 1
Tips
Monday $10.35
Wednesday 7.50
Thursday 6.50
Friday 5.20
Saturday 15.80

Week 2
Wednesday $6.35
Thursday 12.75
Friday 20.35

Week 3
Monday $5.00
Wednesday 7.20
Friday 16.30

Week 4
Tuesday $7.25
Thursday 4.85
Sunday 17.25

Checking the Phone Bill

Checking the phone service bill is a good way to practice your addition and subtraction skills.

```
MONTHLY  SERVICE  8-10  THRU  9-9  (INCL    120  UNITS)  1320
    123  UNITS USED/    120 ALLOWED/      3 ADDL  BILLED    17
                                                          25CR
ITEMIZED CALLS - SEE DETAIL                               1861

    U.S.TAX .96        STATE* .29                          125

                                       TOTAL  DUE  3298
```

1. The monthly service bill states that the customer owes $32.98.

2. To decide if this bill is correct, add the monthly service charge, the extra units billed, the itemized calls, and the U.S. and state taxes.

$13.20 + $.17 + $18.61 + $1.25 = $33.23

(The U.S. tax and state tax are added to get $1.25.)

3. Look at the service bill again. A 25CR is listed. This is a $.25 credit to the customer carried over from the last bill. The credit should be subtracted from the amount owed.

$33.23 − $.25 = $32.98

The total due is $32.98. The bill is correct.

Find the total due on these phone bills.

1.
```
MONTHLY SERVICE 11-10 THRU 12-9  (INCL    120 UNITS) 1320
    89 UNITS USED/    120 ALLOWED/     0 ADDL BILLED
                                                      25CR
ITEMIZED CALLS - SEE DETAIL                           1197
    U.S.TAX .75        STATE* .26                      101

                                   TOTAL  DUE
```
1._____

2.
```
MONTHLY SERVICE 7-10 THRU 8-9  (INCL    120 UNITS) 1320
    166 UNITS USED/    120 ALLOWED/     46 ADDL BILLED 265
                                                      25CR
ITEMIZED CALLS - SEE DETAIL                           251
    U.S.TAX .55        STATE* .31                      86

                                   TOTAL  DUE
```
2._____

Subtracting Decimals

Subtracting decimals is like subtracting whole numbers. To subtract decimals, follow these steps:

Step 1 Write the numbers in a column, lining up the decimal points. Put the larger number on the top.

Step 2 If necessary, add on zeros as placeholders.

Step 3 Subtract as with whole numbers.

Step 4 Bring the decimal point straight down into the answer.

Examples

A. $35.72 - 11.11 = ?$

$$\left.\begin{array}{r} 3\,5\,.7\,2 \\ -\,1\,1\,.1\,1 \end{array}\right] \text{Step 1}$$

$$\left.\begin{array}{r} 3\,5\,.7\,2 \\ -\,1\,1\,.1\,1 \\ \hline 2\,4\quad 6\,1 \end{array}\right] \text{Step 3}$$

$$\begin{array}{r} 3\,5\,.7\,2 \\ -\,1\,1\,.1\,1 \\ \hline 2\,4\,.6\,1 \end{array}$$
↑——— Step 4

B. $.5 - .007 = ?$

$$\left.\begin{array}{r} .5 \\ -\,.0\,0\,7 \end{array}\right] \text{Step 1}$$

$$\left.\begin{array}{r} .5\,0\,0 \\ -\,.0\,0\,7 \end{array}\right] \text{Step 2}$$

$$\left.\begin{array}{r} .5\,0\,0 \\ -\,.0\,0\,7 \\ \hline 4\,9\,3 \end{array}\right] \text{Step 3}$$

$$\begin{array}{r} .5\,0\,0 \\ -\,.0\,0\,7 \\ \hline .4\,9\,3 \end{array}$$
↑——— Step 4

Practice

Subtract.

1. $\begin{array}{r} \$3\,3\,7. \\ -\ \ 1\,9\,2\,.2\,4 \end{array}$

2. $\begin{array}{r} 8\,.9\,2 \\ -\ \ .0\,3\,4 \end{array}$

3. $\begin{array}{r} 1\,7\,.6\,2\,5 \\ -\,1\,5\,.8\,1 \end{array}$

4. $\begin{array}{r} \$7\,2\,7\,.7\,5 \\ -\ \ \ \ 7\,.9\,9 \end{array}$

5. 25 − 24.85

_____

6. $249.99 − $55.10

_____

7. .8086 − .6087

_____

8. $ 1 2 6 . 5 4
−   9 . 1 9

9. . 9 8 2 3
− . 5 5 0 5

10. 6 . 2 1 7
−   . 7 1 0 9 9

11. 624.8 − 1.578

_____

12. .555 − .05

_____

13. 1327.82 − 131.579

_____

14. 129.7 − 64.43

_____

15. .0962 − .0015

_____

16. 365.24 − 78.0897

_____

17. .017 − .009

_____

18. $455.90 − $255.85

_____

19. 23.69 − 6.2495

_____

20. 9 7 2 . 5
−   8 . 4 7 6 1

21. 4 . 3 5
− 1 . 7 5 1

22. 7 . 6
− 1 . 3 7 8 9

23. . 0 4 9 2
− . 0 3 6 7 1

24. 7 . 2
− 0 . 7 5 1

25. 8 . 1 0 4
− 5 .

26. $ 5 4 7 .
−   9 8 . 6 7

27. 5 . 3 8 1
− 3 . 9 9 9 9

28. 9 9 5 . 7 0 1 9
− 3 8 8 . 4 3 6 4

29. Find the difference between .0031 and .010.

_____

30. Subtract $572.82 from $755.41.

_____

45

Solve the following problems.

31. Gregg traveled 321.7 miles on Monday. Peter traveled 155.9 miles on the same day. How many more miles did Gregg travel than Peter?

_____

32. Luke had a charge balance at the Uppermost Clothing Store of $59.29. He made a payment of $23.50. What is his new balance?

_____

33. Christine ran the 100-yard dash in 9.8 seconds. Juanita ran it in 10.3 seconds. Christine was how much faster than Juanita?

_____

34. Teri spent $7.39 at the store. She gave the clerk a $20 bill. How much change did she get?

_____

35. Eduardo had a metal pipe 12.5 feet long. He cut off 4.9 feet. How much pipe was left? (He lost nothing in the cutting.)

_____

36. A vacuum cleaner selling for $125 is on sale for $89.85. If Alan buys the vacuum cleaner now, how much will he save?

_____

37. Shasha earned $199.89 this week. Last week she earned $157.39. How much more did she earn this week than last week?

_____

38. Ted had $132.50 in his checking account. He wrote checks for $15.75, $13.26, and $29.89. How much money did he have left in his checking account?

_____

LIFE SKILL

Making change

The usual policy for making change is to give the customer as few coins as possible.

You pay $ 1 . 0 0
Your bill is   −   . 8 9
Your change   . 1 1

Most likely, you will get one dime and one penny in change.

Circle the correct answer. Use the fewest coins as possible.

1. Stanley's laundry was $7.35. He gave the clerk a $20 bill. How much did he receive in change?

   A. ①①①①① ⑩ ㉕㉕ [1] [1] [5] [5]

   B. ⑤ ⑩ ㉕㉕ [1] [1] [10]

2. Nori gave the clerk $50. How much change should she receive if her bill was $15.56?

   A. [10] [10] [10] [1] [1] [1] [1]
   ⑩⑩⑩⑩ ①①①①

   B. [20] [10] [1] [1] [1] [1] ㉕ ⑩ ⑤ ①①①①

Find the amount of change. Then list the money the customer will most likely receive.

3. David paid $ 9 0 . 0 0   **pennies** _____   **one-dollar bills** _____
   His bill was   −   8 1 . 7 8

                                 nickels _____   **five-dollar bills** _____

   His change   _____

                                 dimes _____   **ten-dollar bills** _____

                                 quarters _____

Problem Solving—Adding and Subtracting Decimals

You have been solving word problems for the previous lessons. You should remember that these problems can be solved if you keep in mind the following steps:

Step 1   Read the problem and underline the key words. These words will generally relate to some mathematics reasoning computation.

Step 2   Make a plan to solve the problem. Ask yourself, Should I add, subtract, multiply, divide, round, or compare? You may have to do more than one of these operations for the same problem.

Step 3   Find the solution. Use your math knowledge to find your answer.

Step 4   Check the answer. Ask yourself, is this answer reasonable? Did you find what you were asked for?

Here are some key words for addition and subtraction.

Addition		**Subtraction**	
altogether	sum	decreased by	how much less
both	total	diminished by	how much more
increase	together	difference	remainder

Examples

A. One yard is 36 inches long. A meter is 39.37 inches long. What is the difference in length between a yard and a meter?

Step 1   Determine the difference between the length of a yard and a meter. The key word is **difference.**

Step 2   The key word indicates which operation should occur—**subtraction.** Before you can perform this operation, you must also determine which number is **greater in value** on the number line.

$$39.37 > 36$$

$$\begin{array}{r} 39.37 \\ -36.00 \\ \hline 3.37 \end{array}$$

Step 3   Find the solution.

Step 4   Check the answer. Does it make sense that 3.37 is the difference between 39.37 and 36. Yes, the answer is reasonable.

B. Mikey purchased a one-way train ticket at the station for $1.55. The return ticket was purchased on the train for $2.15. What was the total cost of the train tickets?

Step 1   Determine the total cost of the train tickets. The key word is **total.**

Step 2   The key word indicates which operation should occur—**addition.**

Step 3   Find the solution.

$$\begin{array}{r} \$1.55 \\ +\ 2.15 \\ \hline \$3.70 \end{array} \quad \text{total cost}$$

Step 4   Check the answer. Does it make sense that $3.70 is the total of $1.55 and $2.15. Yes, the answer is reasonable.

Practice

Problem Solving.

Solve the following, using the steps for solving word problems.

1. A car's odometer read 84,531.8 miles at the beginning of a trip. At the end of the trip, it read 87,675.4 miles. What was the total distance of the trip?   _____

2. Ashley earns $324 biweekly before deductions. If her take home pay is $128.25 per week, how much are the deductions?   _____

3. Combined Charities has a goal of $5.3 million for this year. To date, $2.75 million has been collected. How much more is needed to reach the goal?   _____

4. Sonya gave the clerk $26.19, which was the exact cost and tax for items purchased at the variety store. If the items purchased were $24.94, what was the tax amount?   _____

5. Liam's grocery bill is $15.65, and he gives the clerk $2.25 in coupons. If Liam gives the clerk $20, how much change should he get?   _____

LIFE SKILL

Depositing in a Checking Account

When you fill in a checking account deposit ticket, remember:
currency means paper money
coin means pennies, nickels, dimes, etc.
less cash received is the amount of money you keep
net deposit is the actual amount deposited to your account

Fill out these deposit slips using the following information and find the net deposit.

1. **Coin**   **Checks**
   50 pennies $15.03
   17 nickels $50.00
   20 dimes $143.73
   25 quarters

                     Less cash
   Currency   **received**
   $25.00 $130.00

2. **Ken Yamada has**
   79 pennies, 20 nickels, 11 dimes,
   4 quarters, and $157.00 in currency
   to deposit.
   He also is depositing a check for
   $59.65 and one for $28.09. He wants
   $200.00 in cash.

3. Bring in a checking account deposit
   ticket from a bank near you. Fill it out
   with the following information.

   Coin: 14 quarters, 10 dimes,
   23 nickels
   Currency: $25
   Checks: $34.37, $65.00, $146.75
   Less cash received: $215.96

   Net deposit: _____

1.

CASH	CURRENCY		
	COIN		
LIST CHECKS SINGLY			
TOTAL FROM OTHER SIDE			
TOTAL			
LESS CASH RECEIVED			
NET DEPOSIT			

2.

CASH	CURRENCY		
	COIN		
LIST CHECKS SINGLY			
TOTAL FROM OTHER SIDE			
TOTAL			
LESS CASH RECEIVED			
NET DEPOSIT			

Four of the following eight deposit tickets have incorrect totals.
Find the mistakes and correct them.

1.

CASH	CURRENCY	47	00
	COIN	38	16
LIST CHECKS SINGLY		52	17
		61	87
		4	15
TOTAL FROM OTHER SIDE			
TOTAL		203	35
▶ LESS CASH RECEIVED			
NET DEPOSIT			

2.

CASH	CURRENCY	27	00
	COIN	54	79
LIST CHECKS SINGLY		101	19
		202	64
		518	29
TOTAL FROM OTHER SIDE			
TOTAL		930	91
▶ LESS CASH RECEIVED			
NET DEPOSIT			

3.

CASH	CURRENCY	52	00
	COIN	6	21
LIST CHECKS SINGLY		8	19
		34	67
		58	19
TOTAL FROM OTHER SIDE			
TOTAL		359	62
▶ LESS CASH RECEIVED			
NET DEPOSIT			

4.

CASH	CURRENCY	89	00
	COIN		
LIST CHECKS SINGLY		36	12
		62	59
		138	42
TOTAL FROM OTHER SIDE		73	12
TOTAL		399	25
▶ LESS CASH RECEIVED			
NET DEPOSIT			

5.

CASH	CURRENCY		
	COIN		
LIST CHECKS SINGLY		571	00
		47	39
		51	28
TOTAL FROM OTHER SIDE		96	82
TOTAL		766	49
▶ LESS CASH RECEIVED			
NET DEPOSIT			

6.

CASH	CURRENCY	25	00
	COIN	138	16
LIST CHECKS SINGLY		4,196	57
		27	53
		112	12
TOTAL FROM OTHER SIDE			
TOTAL		5,498	83
▶ LESS CASH RECEIVED			
NET DEPOSIT			

7.

CASH	CURRENCY	116	00
	COIN	97	15
LIST CHECKS SINGLY		16	34
		109	82
		9	25
TOTAL FROM OTHER SIDE			
TOTAL		249	50
▶ LESS CASH RECEIVED			
NET DEPOSIT			

8.

CASH	CURRENCY	201	00
	COIN	76	31
LIST CHECKS SINGLY		175	49
		258	53
		29	99
TOTAL FROM OTHER SIDE			
TOTAL		741	32
▶ LESS CASH RECEIVED			
NET DEPOSIT			

Adding and Subtracting Decimals Using a Calculator

Using a calculator can cut down the time it takes to find the answers to math problems. There are many kinds of calculators on the market today. Most calculators have at least the keys shown here.

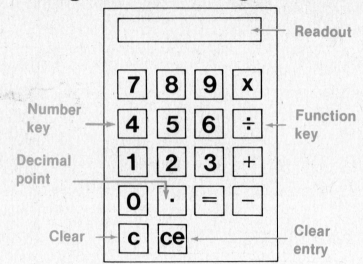

The **readout** window, at the top, is where the answer shows.
The **number keys,** at the left, are for punching in the numbers that you're working the problem with.
The **function keys,** at the right, tell the calculator what to do.

$+$ is used to **add.**

$-$ is used to **subtract.**

$\times$ is used to **multiply.**

$\div$ is used to **divide.**

$=$ means **equals.**   Press this key when you've told the calculator what to do and you want the answer.

$\boxed{C}$ means **clear.**   When you press it, everything in the calculator is erased and you can start over.

$\boxed{ce}$ means **clear entry.**   When you press it, you erase only the last number you put into the calculator. Not all calculators have this key.

$\boxed{\cdot}$ is for **decimals.**   When you press it, the number after it is a decimal.

You can use your calculator to add or subtract decimals.

A. Add. 3.9825 + 6.09

Step 1   First, press the $\boxed{\text{C}}$ button. This will clear out the last problem.

Step 2   Press $\boxed{3}\;\boxed{\cdot}\;\boxed{9}\;\boxed{8}\;\boxed{2}\;\boxed{5}$.

Step 3   Press $\boxed{+}$.

Step 4   Press $\boxed{6}\;\boxed{\cdot}\;\boxed{0}\;\boxed{9}$.

Step 5   Press $\boxed{=}$. Read the answer and write it down.
10.0725

Step 6   Press $\boxed{\text{C}}$ before starting the next problem.

B. Subtract. .781 − .5307

Step 1   First, press the $\boxed{\text{C}}$ button. This will clear out the last problem.

Step 2   Press $\boxed{\cdot}\;\boxed{7}\;\boxed{8}\;\boxed{1}$.

Step 3   Press $\boxed{-}$.

Step 4   Press $\boxed{\cdot}\;\boxed{5}\;\boxed{3}\;\boxed{0}\;\boxed{7}$.
The calculator will automatically adjust .781 to four places.

Step 5   Press $\boxed{=}$. Read the answer and write it down.
.2503

Step 6   Press $\boxed{\text{C}}$ before starting the next problem.

Practice

Add or subtract.

1.
```
      . 0 6
    4 . 8 3 1
  2 5 . 0 3
+     . 4 9 9 8
```

2.
```
  3 5 . 0 7 1 5
  6 7 . 0 3 4
  7 7 . 0 9
+     . 6 7 3
```

3.
```
  . 7 2 3 9
  . 5 4 7 2
  . 3 1 6 1
+ . 8 6 5 9 2
```

4.
```
  3 9 7 . 2 1
−  4 8 . 0 6 7
```

5.
```
  2 0 .
−     . 0 2
```

6.
```
  . 5 1 3
− . 3 2 1 7 8
```

LIFE SKILL

Calculators/Buying a New Car

Mr. and Mrs. Anders want to buy a new car. Mr. Anders wants a mid-size car. Mrs. Anders wants a compact car. They discuss this major purchase and agree that they will not spend more than $9,500 after their trade-in allowance of $2,500. The $9,500 will not include the financing or the taxes.

Below are the stickers for the two cars they selected. Find the total cost for each car. Which car did the Anders most likely buy? _____

Mr. Anders' choice

Empire	
BASE PRICE	$8,649.53
Custom Shoulder and Seat Belts	42.00
Tinted Glass	170.00
Landau Top	528.00
Electric R/W Defogger	199.00
Air Conditioner	662.00
Outside R/V Sport Mirror	40.00
Engine-V8 301 Cu. In.	810.00
Automatic Trans.	505.00
Power Steering	363.00
Steel Belted Radial W/W Tires	640.00
Convenience Group	119.00
AM-FM Radio	163.00
Exterior Molding	153.00
Transportation	178.00
Total	_____

Mrs. Anders' choice

Bluebird	
BASE PRICE	$7,950.00
Custom Shoulder & Seat Belts	28.00
Tinted Glass	160.00
Roof Crown Molding	183.00
Floor Mats	48.00
Door Guards	22.00
Electric R/W Defogger	696.00
Air Conditioner	500.00
Power Front Disc Brakes	650.00
Automatic Trans.	68.00
Tilt Steering Wheel	146.00
Power Steering	430.00
Four Styled Aluminum Wheels	436.00
Steel Belted Radial W/S Tires	148.00
AM-FM Stereo Radio	24.00
Convenience Group	73.00
Instrument Gauges	164.00
Transportation	100.00
Total	_____

Go to two different car dealers. Compare the sticker prices of the same make of compact car. What prices or options are different?

Posttest

Find the sums and differences for the following.

1. $10.3 + 3.45 + 234.972$ _____

2. $5.098 + 5.1 + 498.8$ _____

3. $1.07 + 6.8471 + 2.17$ _____

4. $4.251 + .0198 + 14.32$ _____

5. $39 - .05067$

6. $7.3 - 2.67$

7. $\$16.06 - \$9.67$

_____

_____

_____

8. $12.1206 - 5.307$

9. $5{,}129 - 4{,}567.89$

10. $52.27 - 39.0057$

_____

_____

Solve the following problems.

11. Mrs. Allen bought groceries for $6.38 and gave the clerk $20. How much change did she get?

_____

12. A city's snowfall was .89 inches in November, 3.96 in December, and 4.073 in January. What was the total snowfall for the three months?

_____

13. Kent walked 4.9 miles on Wednesday. On Friday, he walked 6.1 miles. How many more miles did he walk on Friday than on Wednesday?

_____

14. After Lillian had saved $17.98, she still needed $8.00 to pay for the blouse that she wanted. What was the price of the blouse?

_____

Use a calculator to solve the following.

15. Armando bought 5.6 pounds of mixed nuts. If he gave 1.6 pounds to Amanda, 1.85 pounds to Michelle, and one pound to Arthur, how many pounds did he have left?

_____

4

Multiplying and Dividing Decimals

Multiply.

1. 0.8×2 _____

2. 227×6.5 _____

3. $.006 \times .40$ _____

4.
$$\begin{array}{r} 0.05 \\ \times\, 0.45 \\ \hline \end{array}$$

5.
$$\begin{array}{r} 5.910 \\ \times\ \ .151 \\ \hline \end{array}$$

6.
$$\begin{array}{r} 11.28 \\ \times\ \ \ \ .45 \\ \hline \end{array}$$

7. 4.607×10 _____

8. 4.607×100 _____

9. 4.607×1000 _____

Divide.

10. $75\overline{)82.5}$

11. $12\overline{)8.340}$

12. $.15\overline{).1860}$

13. $.24\overline{).0168}$

14. $.532\overline{).08512}$

15. $3.01\overline{)653.17}$

16. $460.7 \div 10$

_____

17. $460.7 \div 100$

_____

18. $460.7 \div 1000$

_____

Solve the following problems.

19. Hand towels are on sale at five for $7. The regular price is $1.99 each. If Mya buys the five on sale, how much will she save?

_____

20. A group of 11 businessmen ordered luggage tags. The total bill was $47. What was the cost per person? Round to the nearest dollar.

_____

21. Roll insulation sells for $10.95 a bundle. How much will 10 bundles cost?

_____

22. Gilbert earns $346.39 a week. His wife earns $221.78 a week. Assuming each works 50 weeks per year, what is their annual income?

_____

Counting Decimal Places

Counting the number of decimal places means counting all the digits to the right of the decimal point. For example, 156.15 has two digits to the right of the decimal point.

Examples

A. 51.**072**
has 3 decimal places.

B. 356.**0001**
has 4 decimal places.

C. 5346.
has 0 decimal places.

D. 728.**3**
has 1 decimal place.

Practice

How many decimal places are in each number?

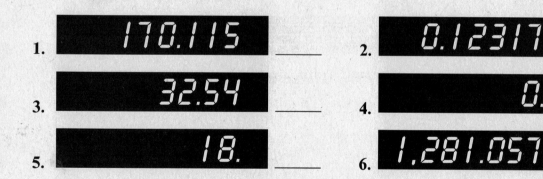

1. 170.115 _____

2. 0.123173 _____

3. 32.54 _____

4. 0.5 _____

5. 18. _____

6. 1,281.0571 _____

Change the numbers below by adding decimal points. Show two decimal places, three decimal places, and four decimal places. The first problem is completed for you.

	2 places	3 places	4 places
7. 16971	169.71	16.971	1.6971
8. 698305			
9. 54173			
10. 18963			
11. 418327			

Multiplying Decimals

Examples

When multiplying decimals, follow these steps:

Step 1   Multiply as with whole numbers.

Step 2   Count the number of decimal places in both numbers of the problem and add them together.

Step 3   Count off this total number of decimal places in the answer. Move the decimal point from right to left.

MATH HINT

When there are fewer digits in the answer than the total number of places to the right of the decimal point, add zeros to the left as placeholders.

A.   Multiply. $23.71 \times 1.5 = ?$

Step 1
$$\begin{array}{r} 2\,3\,.\,7\,1 \\ \times\ \ \ \ 1\,.\,5 \\ \hline 1\,1\,8\,5\,5 \\ 2\,3\,7\,1\ \ \\ \hline 3\,5\,5\,6\,5 \end{array}$$

Step 2
$$\begin{array}{r} 2\,3\,.\,7\,1 \\ \times\ \ \ \ 1\,.\,5 \end{array}$$
$\quad$ 2 places
$+$ 1 place
$\overline{\quad}$ 3 places

Step 3
$$\begin{array}{r} 2\,3\,.\,7\,1 \\ \times\ \ \ \ 1\,.\,5 \\ \hline 3\,5\,.\,5\,6\,5 \end{array}$$
$\quad$ 3 places

The answer is 35.565.

B.   Multiply. $7.09 \times .003 = ?$

Step 1
$$\begin{array}{r} 7\,.\,0\,9 \\ \times\,.\,0\,0\,3 \\ \hline 2\,1\,2\,7 \end{array}$$

Step 2
$$\begin{array}{r} 7\,.\,0\,9 \\ \times\,.\,0\,0\,3 \end{array}$$
$\quad$ 2 places
$+$ 3 places
$\overline{\quad}$ 5 places

$$\begin{array}{r} 7\,.\,0\,9 \\ \times\,.\,0\,0\,3 \\ \hline 0\,2\,1\,2\,7 \end{array}$$

Step 3
$$\begin{array}{r} 7\,.\,0\,9 \\ \times\,.\,0\,0\,3 \\ \hline .\,0\,2\,1\,2\,7 \end{array}$$
$\quad$ 5 places

The answer is .02127.

Multiply.

1. 14.21
    ×   35

2. .216
    × .24

3. 17.7
    × 2.5

4. 3.2
    × 87

5. 547.1
    ×   .73

6. 3.07
    × 731

7. 17.39
    × 12.4

8. 1.754
    ×   .23

9. .02
    × .8

10. .25
     ×.04

11. 1.2
     ×.008

12. .073
     × .05

13. 4.08
     ×.0017

14. .0125
     × .315

15. .468
     ×.1302

16. .12121
     ×   .03

Solve the following problems. The first one is done for you.

17. The Valdez car goes 18.2 miles on one gallon of gasoline. How far can it go on 16.5 gallons of gas?

300.30 miles

          18.2   1 place
        × 16.5   + 1 place
          910   2 places
         1092
          182
        300.30

18. One page in the telephone book is .0057 inch thick. The book has 1172 pages. How thick is the book?

_____

19. One gallon of milk sells for $1.98. If the Hall family drinks a gallon a day, how much will they spend for the week?

_____

20. Ramona bought 11 yards of curtain material. The material cost $4.98 a yard. How much did she spend?

_____

21. Lin walked 1.5 blocks to the train station 5 times a week for 50 weeks. How far did she walk in the 50 weeks, round trip?

_____

22. If you can save $15.75 each month, how much will you save in two years?

_____

23. Alex bought 4 spark plugs for $1.20 each and an oil filter for $3.50. How much did he spend for these items?

_____

24. Edith bought 5 cans of pears at $.69 a can, 3 pounds of onions at $.29 a pound, and a pound of ham at $1.88 a pound. How much change will she get from $10?

_____

25. Mr. Nogales averaged 47.9 miles per hour for 7 hours. How far did he drive?

_____

LIFE SKILL

Reconciling a Bank Statement

To reconcile the bank statement means to make sure that the checking account statement agrees with your checkbook balance. Sometimes the balances do not agree. This happens when you have **outstanding checks**—checks you have written which had not reached the bank for payment before the statement was prepared—and deposits that appear in your checkbook but not on the bank statement. It can also happen when there are bank service charges that have not been entered in your checkbook.

For example, Mary Carter's bank statement shows a balance of $286.30. Her checkbook balance is $279.38. She had made a deposit of $52.70, which does not appear on her bank statement. She has three outstanding checks: $43.93; $100.00; and $16.00. The bank statement also shows service charges of $2.45; $58.36; and $39.50.

Let's help Mary reconcile her statement by completing the following steps.

The amounts on Lines 4 and 8 should be the same. Remember, reconcile means "to be the same" or "to be in agreement."

CHECKING ACCOUNT STATEMENT

Balance shown on bank statement	$286.30
Add deposits not on statement	$ 52.70
Subtotal	$ ___ ①
Outstanding Checks	___ ②
	___
	___
Total Outstanding Checks	___ ③
Balance	___ ④

Step 1 Add deposits made but not shown on Mary's bank statement.

Step 2 List all outstanding checks in the spaces beside and below Line 2.

Step 3 Add the outstanding checks. Write the total on Line 3.

Step 4 Subtract Line 3 from Line 1. Write your answer on Line 4.

MARY'S CHECKBOOK BALANCE

Balance in Mary's checkbook	$279.38
Deposits not entered in Mary's checkbook	0
Subtotal	___ ⑤
Bank Service charges	___
	___ ⑥
	___
Total charges	___ ⑦
Balance	___ ⑧

Step 5 Add any deposits not entered in Mary's checkbook to her checkbook balance. Write the total on Line 5.

Step 6 List bank service charges that were not entered in the checkbook. List them in the spaces beside Line 6.

Step 7 Add the charges. Write the total on Line 7.

Step 8 Subtract Line 7 from Line 5. Write the answer on Line 8.

Find the balance forward at the bottom of each check stub.

9.

	DOLLARS	CENTS
BALANCE BROUGHT FORWARD	215	36
ADD DEPOSITS	45	00
TOTAL	260	36
LESS THIS CHECK	79	84
LESS OTHER DEDUCTIONS	3	00
BALANCE CARRIED FORWARD	177	52

10.

	DOLLARS	CENTS
BALANCE BROUGHT FORWARD	21	65
ADD DEPOSITS	782	06
TOTAL	803	71
LESS THIS CHECK	521	54
LESS OTHER DEDUCTIONS		
BALANCE CARRIED FORWARD		

11.

	DOLLARS	CENTS
BALANCE BROUGHT FORWARD	116	75
ADD DEPOSITS	47	35
	41	22
	127	99
TOTAL	333	31
LESS THIS CHECK	159	76
LESS OTHER DEDUCTIONS	15	09
BALANCE CARRIED FORWARD		

12.

	DOLLARS	CENTS
BALANCE BROUGHT FORWARD		
ADD DEPOSITS	115	79
	237	81
	432	11
TOTAL	785	71
LESS THIS CHECK		
LESS OTHER DEDUCTIONS		
BALANCE CARRIED FORWARD		

Find the balance at the bottom of each check stub. This balance becomes the balance forward at the top of the next check stub.

13.

	DOLLARS	CENTS
BALANCE BROUGHT FORWARD	547	16
ADD DEPOSITS	47	25
TOTAL	594	41
LESS THIS CHECK	234	89
LESS OTHER DEDUCTIONS		
BALANCE CARRIED FORWARD		

14.

	DOLLARS	CENTS
BALANCE BROUGHT FORWARD		
ADD DEPOSITS	25	15
TOTAL	384	67
LESS THIS CHECK	117	41
LESS OTHER DEDUCTIONS	11	00
BALANCE CARRIED FORWARD		

Multiplying Decimals by 10, 100, or 1,000

There is a shortcut for multiplying a decimal by 10, 100, or 1,000.
Note the number of zeros in these numbers. In 10, there is one zero.
In 100, there are two zeros, and in 1,000, there are three zeros. When
multiplying a decimal by these numbers, move the decimal point **to
the right** as many places as there are zeros. Add zeros if necessary.

Examples

A. $10 \times 27.5 = 27.5 = 275$

$10 \times 2.75 = 2.7.5 = 27.5$

$10 \times .275 = .2.75 = 2.75$

B. $100 \times 1.53 = 1.53. = 153$

$100 \times 15.3 = 15.30. = 1,530$

$100 \times 153. = 153.00. = 15,300$

C. $1,000 \times 1.675 = 1.675. = 1,675$

$1,000 \times 16.75 = 16.750. = 16,750$

$1,000 \times 167.5 = 167.500. = 167,500$

D. $10 \times 2,571.5. = 25,715$

$100 \times 2,571.50. = 257,150$

$1,000 \times 2,571.500. = 2,571,500$

Practice

Multiply.

1. 10×7.6

2. $100 \times .013$

3. 10×7.64

4. $10 \times .013$

5. $10 \times .7839$

6. $100 \times .002$

7. $100 \times .7839$

8. $10 \times .9084$

9. $1,000 \times .7839$

10. $1,000 \times 908.4$

11. 10×7.9

12. 100×90.84

Completing Supply Orders

Complete the supply order for John's Ice Cream Stand and the Kut and Kurl. The first item is done for you.

Cost/per means the unit cost for each item given. Multiply to find the total cost.

	Supply Order		**John's Ice Cream Stand**	
	Item	Quantity	Cost/per	Total
1.	Ice Cream	16 cartons	$4.19 per carton	$ 67.04
2.	Cones	20 boxes	$1.39 per box	_____
3.	Syrup	3 gal.	$3.29 per gal.	_____
4.	Nuts	5 lbs.	$4.31 per lb.	_____
5.	Whipped Cream	12 cans	$ .59 per can	_____
6.	Bananas	12 lbs.	$ .29 per lb.	_____
7.	Cherries	2 qt.	$1.59 per qt.	_____
8.	Total			_____

(handwritten calculation)
$4.19
× 16
2514
419
$ 67.04

	CUSTOMER *Kut and Kurl 13215 Wood St. Blue Island, Illinois 60421*			**Shear Madness Beauty Supplies** 15932 Ashland Ave. Harvey, Illinois 60426	
	Item	Order No.	Quantity	Price for one	Total Price
9.	Nail Care Kit	662-464	10	6.99	_____
10.	Facial Care Set	662-452	2	8.99	_____
11.	Hair Conditioner	763-921	25	1.79	_____
12.	Hair Shampoo	675-123	30	1.59	_____
13.	Budget Haircutting Kit	205-421	1	12.99	_____
14.	Total				_____

Dividing a Decimal by a Whole Number

Division means to divide or separate a number into equal groups. Read the example below. It asks you to divide the cost of a car into 30 equal payments.

$$\overset{\text{quotient}}{\text{divisor}\overline{)\text{dividend}}}$$

Example

The total cost of a new car is $8,286 to be paid in 30 monthly payments. What is each monthly payment?

To do decimal division, follow these steps:

Step 1 Place a decimal point after the last digit of the dividend, if it is a whole number.

Step 2 Place a decimal point in the quotient directly above the decimal point in the dividend.

Step 3 Divide as with whole numbers.

Step 4 Add zeros after the decimal point of a whole-number dividend, if there is a remainder. This way, the remainder will be written as a decimal amount.

```
                ↓──────── Step 2
      $ 276.20
   30)$8286.00  ←──── Step 3
     −60    ↑──────── Step 1
      228
     −210
      186
     −180
       6 0
      −6 0
        00
        00
```

6 is the remainder, written as the decimal amount .20 in the quotient.

Practice

Divide.

1. $4\overline{)33.12}$ **2.** $9\overline{)2.844}$

3. $14\overline{)52.626}$ **4.** $24\overline{)6.072}$

5. $275.9 \div 31$ **6.** $72 \div 16$

_____   _____

7. $2,983.2 \div 66$ **8.** $637 \div 14$

_____   _____

Comparing Meal Costs

Mrs. Allen was planning meals for the week for her family of six.
Which of the meals is the most expensive in terms of cost per person?
Which meal is the least expensive in terms of cost per person?

To find the cost per person:
1. Add all the item prices that make up the meal.
2. Divide the total cost by the number of people eating.

Roast beef dinner
at home
Roast beef $ 6.50
Dinner rolls   1.59
Can corn   .79
Salad   1.79
Can green beans   .79
Cake as dessert   3.49
½ gal. milk   1.59
Tax   1.28

Total

Cost per person

Roast beef dinner
at restaurant
6 roast beef dinners
  at $6.50 each
6 glasses of milk
  at $.85 each
6 slices of cake
  at $1.25 each

Roast beef dinners
included corn, green
beans, salad and rolls
Tax   3.18
Tip   6.00

Total

Cost per person

Joe's Pizza
1 jumbo pizza $ 9.50
1 2-liter
  soft drink   1.59
Delivery
  charge   .50
Tax   .88

Total

Cost per person

Fast food
6 hamburgers at
  95¢ each
6 orders of fries at
  45¢ each
6 soft drinks at
  35¢ each   1.16
Tax

Total

Cost per person

Dividing a Decimal by a Decimal

To divide a decimal by a decimal follow these steps:

Step 1 Make the divisor a whole number by moving the decimal point to the right of the last digit.

Step 2 In the dividend, move the decimal point to the right the same number of places.

Step 3 Now place a decimal point directly above in the quotient.

Step 4 Divide as with whole numbers.

Example

Find the quotients.

A. $7.896 \div .24 = ?$

Step 1 $.24.\overline{)7.896}$

Step 2 $24.\overline{)7.89.6}$

Step 3 $24.\overline{)789.6}$

Step 4
$$
\begin{array}{r}
32.9 \\
24.\overline{)789.6} \\
-72 \\
\hline
69 \\
-48 \\
\hline
21\,6 \\
-21\,6 \\
\hline
\end{array}
$$

B. $0.2398 \div .011 = ?$

Step 1 $.011.\overline{)0.2398}$

Step 2 $11\overline{)0.239.8}$

Step 3 $11\overline{)0239.8}$

Step 4
$$
\begin{array}{r}
21.8 \\
11\overline{)0239.8} \\
-22 \\
\hline
19 \\
-11 \\
\hline
8\,8 \\
-8\,8 \\
\hline
\end{array}
$$

Find the quotients. Show your work.

1. $.03\overline{)2.493}$

2. $48.2\overline{)77.12}$

3. $8.3\overline{)1.245}$

4. $.019\overline{)0.855}$

5. $4.8\overline{)393.60}$

6. $2.10\overline{)1.281}$

7. $9.4\overline{)9.682}$

8. $.7\overline{)4.354}$

9. $.0004\overline{)2648}$

Problem Solving

Solve the following problems.

10. Mr. Allison has $26.25 to buy basketball play books for his team. If each book costs $2.25, how many books can he buy?

_____

11. A metal rod is 44.24 inches long. If the rod is divided into four equal pieces, how long is each piece?

_____

12. Mrs. Ngo had 10.5 pounds of strawberries. She divided them into packages of .75 pounds. How many packages did she have?

_____

13. A developer divided 11.25 acres of land into .25 acre lots. How many lots did he make?

_____

LIFE SKILL

Eating Out

WALNUT INN

Soup

Split pea	Bowl	.95
Vegetable beef	Bowl	.95

Dinner Suggestions

Top Sirloin Steak
Only the most tender part of the
sirloin is used for this steak ... 7.25

Double Lamb Chops
Double treat in store for you with
this tender, juicy delight ... 7.25

Roast Long Island Duckling
Served with cumberland sauce and
cornbread dressing ... 4.25

**Fried Calves' Liver with Grilled
Onions** ... 4.40

Fried Chicken, New Orleans Style
One-half spring chicken, batter
fried to a golden brown ... 5.75

Child's Plate
Hamburger steak or shrimp or fillet
of sole, vegetables and potatoes ... 3.50

Dessert

Apple pie	1.25
Chocolate cake	1.25
Ice cream (chocolate or vanilla)	1.25

Beverages

Coffee, tea		.60
Milk	.80	1.00
Soft drinks	.80	1.00

The Wilson family had Sunday dinner at the Walnut Inn Restaurant.

Fill in the guest check and determine the subtotal.

GUEST CHECK

TABLE NO.	NO. PERSONS	CHECK NO. 80466	SERVER NO.	
1	Lamb Chops			
2	Child's Plate			
1	Duckling			
1	Sirloin Steak			
1	Fried Chicken			
2	Vegetable Soup			
2	Coffee			
2	Milk, small			
2	Soft drinks, large			
2	Apple Pies			
2	Chocolate Cakes			
2	Ice Cream			
		Subtotal		
TAX				

Dividing Decimals With Zeros in the Answer

When dividing decimals, zeros may be used as placeholders in the tenths and hundredths places. This occurs when the dividend cannot be divided by the divisor. Follow these steps:

Step 1   Move the decimal points to the right.

Step 2   Divide as with whole numbers.

Examples

A. Divide.   .927 ÷ .3 = ?

Step 1 $.3.\overline{).9.27}$

Step 2
```
    3.09
3.)9.27
   −9
    2
   −0
    27
   −27
```

B. Divide.   .00009 ÷ .03 = ?

Step 1 $03.\overline{).00.009}$

Step 2
```
      .003
3.)00.009
   −0
    0
   −0
    9
   −9
```

MATH HINT

It is necessary to use the zero as a placeholder in the tenths place because 3 cannot be divided into 2.

Practice

Divide.

1. $.8\overline{)32.48}$

2. $8\overline{).1640}$

3. $.24\overline{)48.48}$

4. $.07\overline{).003451}$

Dividing Decimals by 10, 100, or 1,000

The shortcut method for multiplying decimals by 10, 100, and 1,000—moving the decimal point—is also a shortcut method for dividing decimals by 10, 100, and 1,000. When dividing, move the decimal point **to the left.**

Examples

A. To divide by 10, move the decimal point **one** place to the left.

$4.63 \div 10 = .4.63 = .463$

$13.2 \div 10 = 1.3.2 = 1.32$

$2,337 \div 10 = 233.7. = 233.7$

B. To divide by 100, move the decimal point **two** places to the left.

$1,275.07 \div 100 = 12.75.07 = 12.7507$

$675.9 \div 100 = 6.75.9 = 6.759$

$.06 \div 100 = .00.06 = .0006$

In the last example above, you had to add two zeros in order to move the decimal point two places to the left.

C. To divide by 1,000, move the decimal point **three** places to the left.

$6,842.8 \div 1,000 = 6.842.8 = 6.8428$

$579.678 \div 1,000 = .579.678 = .579678$

$.1 \div 1,000 = .000.1 = .0001$

In the example at the left, you had to add three zeros in order to move the decimal point three places to the left.

Practice

Divide using the shortcut method.

1. $7.7 \div 10 =$ _____   **2.** $.14 \div 10 =$ _____   **3.** $8.306 \div 100 =$ _____

4. $52.9 \div 100 =$ _____   **5.** $64.7 \div 1,000 =$ _____   **6.** $50.73 \div 1,000 =$ _____

Determining Gross and Net Pay

Dante Amos works for the David Construction Company. He works 40 hours a week at an hourly rate of $8. He receives an overtime rate of $12 for any extra hours he works.

Dante worked 42 hours this week. To determine his **gross pay** for this week, follow these steps.

Step 1	Multiply the number of regular hours worked by the hourly rate.	4 0 regular hours × 8 hourly rate $ 3 2 0 regular earnings
Step 2	Multiply the overtime rate by any overtime hours worked.	$ 1 2 overtime rate 2 hours overtime $ 2 4 overtime earnings
Step 3	Add the regular earnings and the overtime earnings to determine the gross pay.	$ 3 2 0 regular earnings 2 4 overtime earnings $ 3 4 4 gross pay

Payroll terms you should know:

Gross pay   is the total amount a person earned before deductions such as taxes are withheld.

Net pay   is the actual amount of money a person takes home after deductions.

Deductions   are the amounts withheld from a person's gross earnings. These deductions include taxes, bonds, health insurance, credit union, and so on.

FICA   means Federal Insurance Contributions Act. This withheld tax, once called FICA, is now identified as two separate taxes—Social Security tax and Medicare tax.

Federal   is the amount withheld for federal income tax.

State   is the amount withheld for state taxes.

Remember: Gross pay is before deductions, and net pay is the actual amount of money a person takes home after deductions.

To determine **net pay,** follow these steps.

Step 1   Add all the deductions.
Step 2   Subtract the total deductions from gross pay.
Step 3   The remaining amount is net pay.

1. Carla Wilkins also works for the David Construction Company. This week she worked 41 hours. Her hourly rate is $8.35, and her overtime rate is $12.53. What is her gross pay for the week?   _____

2. Dirk Bryant worked 36 hours this week. His hourly rate is $7.50, and his overtime rate is $11.25. What is his gross pay for the week?   _____

3. Kendra Sanchez worked 45 hours this week. Her hourly rate is $10, and her overtime rate is $15. What is her gross pay for the week?   _____

LIFE SKILL

Now look at the following payroll register for the David Construction Company. The deductions for Dante Amos are:

Step 1
$20.32 Social Security Tax (add all the deductions)
  4.75 Medicare Tax
 41.40 Federal
  7.62 State
  5.25 Bonds
$79.34 Total deductions

Step 2
$344.00 gross pay   (subtract deductions from
 79.34 total deductions   the gross pay)
$264.66 net pay

Using the gross pay you found for Carla Wilkins, Dirk Bryant, and Kendra Sanchez, fill in the payroll register. Also determine the total deductions and net pay.

David Construction Company
Payroll Register

Pay Period Ending May 19   19_____   Date of Payment May 19   19_____

| NAME | GROSS PAY | Deductions | | | | | | | NET PAY |
		Soc. Sec. Tax	Medicare Tax	Federal	State	Bonds	Health Ins.	Total	
AMOS, D.	344.00	20.32	4.75	41.40	7.62	5.25	—	79.34	264.66
BRYANT, D.		20.84	5.10	35.20	7.68	2.50	15.00		
SANCHEZ, K.		18.88	3.80	32.90	7.08	7.50	10.00		
WILKINS, C.		9.76	2.75	23.50	3.66	—	—		

Rounding Quotients

Example

To round a quotient to the nearest **tenth,** follow these steps:

Step 1   Before dividing, move decimal points, if necessary. In this case, moving decimal points is not necessary.

$$8\overline{)37.69}$$

Step 2   Divide to one place more than the one to which you are rounding. In this case, divide to hundredths.

$$\begin{array}{r} 4.71 \\ 8\overline{)37.69} \\ -32 \\ \hline 5\,6 \\ -5\,6 \\ \hline 9 \\ -8 \\ \hline 1 \end{array}$$

Step 3   Disregard the remainder.

Step 4   Apply the rules for rounding decimals.   The answer is 4.7.

Practice

Divide. Round each quotient to the nearest tenth.

1. $6\overline{)18.1}$

2. $.45\overline{).1962}$

3. $1.25\overline{)8.13}$

4. $1.4\overline{)7.36}$

5. $3.2\overline{)8.625}$

6. $.16\overline{)1.376}$

7. $1.2\overline{)39.39}$

8. $13\overline{)96.09}$

To round a quotient to the nearest **hundredth,** follow these steps:

Step 1   Before dividing, move decimal points, if necessary.

$$4.9.\overline{)12.4.57}$$

Step 2   Divide to one place more than the one to which you are
rounding. In this case, divide to thousandths.

```
        2.542
49.)124.570
    −98
    ────
     26 5
    −24 5
    ─────
      2 07
     −1 96
     ─────
       110
      − 98
      ────
        12
```

Step 3   Disregard the remainder.

Step 4   Apply the rules for rounding decimals.   The answer is 2.54.

Divide. Round each quotient to the nearest hundredth.

9. $.04\overline{)\,.0579}$ **10.** $.15\overline{)\,.3657}$ **11.** $2.17\overline{)4.326}$

12. $8.4\overline{)12.503}$ **13.** $42\overline{)22.644}$ **14.** $17.9\overline{)\,.8512}$

LIFE SKILL

Adding Sales Tax to Cost of Purchases

In many states, sales tax is charged on purchases of food, clothing, furniture, and other items. To save time for salespeople, tax schedules are printed.

If a customer makes purchases of $10.36, $5, $3.47, and $6.11 in a clothing store, how much does he pay, including tax?

1. Add the purchases.

$$\begin{array}{r} \$10.36 \\ 5.00 \\ 3.47 \\ + \quad 6.11 \\ \hline \$24.94 \end{array} \text{ amount of sale}$$

2. Find the amount of sale on the chart. Read the tax.

3. Add the tax to the amount of sale.

$$\begin{array}{r} \$24.94 \quad \text{amount of sale} \\ + \quad 1.25 \quad \text{tax} \\ \hline \$26.19 \quad \text{total} \end{array}$$

Use the tax schedule. Find the tax and total for these purchases.

1. $.98, $.72, $.31, $1.65, $4.22, and $1.59
Tax _____ Total _____

2. $9.53, $4.26, $1.49, $3.78, and $4.29
Tax _____ Total _____

3. $26.91, $.48, $3.75, $6.01, and $.13
Tax _____ Total _____

Amount of Sale	Tax	Amount of Sale	Tax	Amount of Sale	Tax
.13 thru .25	.01	12.70 " 12.89	.64	25.30 " 25.49	1.27
.26 " .46	.02	12.90 " 13.09	.65	25.50 " 25.69	1.28
.47 " .67	.03	13.10 " 13.29	.66	25.70 " 25.89	1.29
.68 " .88	.04	13.30 " 13.49	.67	25.90 " 26.09	1.30
.89 " 1.09	.05	13.50 " 13.69	.68	26.10 " 26.29	1.31
1.10 " 1.29	.06	13.70 " 13.89	.69	26.30 " 26.49	1.32
1.30 " 1.49	.07	13.90 " 14.09	.70	26.50 " 26.69	1.33
1.50 " 1.69	.08	14.10 " 14.29	.71	26.70 " 26.89	1.34
1.70 " 1.89	.09	14.30 " 14.49	.72	26.90 " 27.09	1.35
1.90 " 2.09	.10	14.50 " 14.69	.73	27.10 " 27.29	1.36
2.10 " 2.29	.11	14.70 " 14.89	.74	27.30 " 27.49	1.37
2.30 " 2.49	.12	14.90 " 15.09	.75	27.50 " 27.69	1.38
2.50 " 2.69	.13	15.10 " 15.29	.76	27.70 " 27.89	1.39
2.70 " 2.89	.14	15.30 " 15.49	.77	27.90 " 28.09	1.40
2.90 " 3.09	.15	15.50 " 15.69	.78	28.10 " 28.29	1.41
3.10 " 3.29	.16	15.70 " 15.89	.79	28.30 " 28.49	1.42
3.30 " 3.49	.17	15.90 " 16.09	.80	28.50 " 28.69	1.43
3.50 " 3.69	.18	16.10 " 16.29	.81	28.70 " 28.89	1.44
3.70 " 3.89	.19	16.30 " 16.49	.82	28.90 " 29.09	1.45
3.90 " 4.09	.20	16.50 " 16.69	.83	29.10 " 29.29	1.46
4.10 " 4.29	.21	16.70 " 16.89	.84	29.30 " 29.49	1.47
4.30 " 4.49	.22	16.90 " 17.09	.85	29.50 " 29.69	1.48
4.50 " 4.69	.23	17.10 " 17.29	.86	29.70 " 29.89	1.49
4.70 " 4.89	.24	17.30 " 17.49	.87	29.90 " 30.09	1.50
4.90 " 5.09	.25	17.50 " 17.69	.88	30.10 " 30.29	1.51
5.10 " 5.29	.26	17.70 " 17.89	.89	30.30 " 30.49	1.52
5.30 " 5.49	.27	17.90 " 18.09	.90	30.50 " 30.69	1.53
5.50 " 5.69	.28	18.10 " 18.29	.91	30.70 " 30.89	1.54
5.70 " 5.89	.29	18.30 " 18.49	.92	30.90 " 31.09	1.55
5.90 " 6.09	.30	18.50 " 18.69	.93	31.10 " 31.29	1.56
6.10 " 6.29	.31	18.70 " 18.89	.94	31.30 " 31.49	1.57
6.30 " 6.49	.32	18.90 " 19.09	.95	31.50 " 31.69	1.58
6.50 " 6.69	.33	19.10 " 19.29	.96	31.70 " 31.89	1.59
6.70 " 6.89	.34	19.30 " 19.49	.97	31.90 " 32.09	1.60
6.90 " 7.09	.35	19.50 " 19.69	.98	32.10 " 32.29	1.61
7.10 " 7.29	.36	19.70 " 19.89	.99	32.30 " 32.49	1.62
7.30 " 7.49	.37	19.90 " 20.09	1.00	32.50 " 32.69	1.63
7.50 " 7.69	.38	20.10 " 20.29	1.01	32.70 " 32.89	1.64
7.70 " 7.89	.39	20.30 " 20.49	1.02	32.90 " 33.09	1.65
7.90 " 8.09	.40	20.50 " 20.69	1.03	33.10 " 33.29	1.66
8.10 " 8.29	.41	20.70 " 20.89	1.04	33.30 " 33.49	1.67
8.30 " 8.49	.42	20.90 " 21.09	1.05	33.50 " 33.69	1.68
8.50 " 8.69	.43	21.10 " 21.29	1.06	33.70 " 33.89	1.69
8.70 " 8.89	.44	21.30 " 21.49	1.07	33.90 " 34.09	1.70
8.90 " 9.09	.45	21.50 " 21.69	1.08	34.10 " 34.29	1.71
9.10 " 9.29	.46	21.70 " 21.89	1.09	34.30 " 34.49	1.72
9.30 " 9.49	.47	21.90 " 22.09	1.10	34.50 " 34.69	1.73
9.50 " 9.69	.48	22.10 " 22.29	1.11	34.70 " 34.89	1.74
9.70 " 9.89	.49	22.30 " 22.49	1.12	34.90 " 35.09	1.75
9.90 " 10.09	.50	22.50 " 22.69	1.13	35.10 " 35.29	1.76
10.10 " 10.29	.51	22.70 " 22.89	1.14	35.30 " 35.49	1.77
10.30 " 10.49	.52	22.90 " 23.09	1.15	35.50 " 35.69	1.78
10.50 " 10.69	.53	23.10 " 23.29	1.16	35.70 " 35.89	1.79
10.70 " 10.89	.54	23.30 " 23.49	1.17	35.90 " 36.09	1.80
10.90 " 11.09	.55	23.50 " 23.69	1.18	36.10 " 36.29	1.81
11.10 " 11.29	.56	23.70 " 23.89	1.19	36.30 " 36.49	1.82
11.30 " 11.49	.57	23.90 " 24.09	1.20	36.50 " 36.69	1.83
11.50 " 11.69	.58	24.10 " 24.29	1.21	36.70 " 36.89	1.84
11.70 " 11.89	.59	24.30 " 24.49	1.22	36.90 " 37.09	1.85
11.90 " 12.09	.60	24.50 " 24.69	1.23	37.10 " 37.29	1.86
12.10 " 12.29	.61	24.70 " 24.89	1.24	37.30 " 37.49	1.87
12.30 " 12.49	.62	24.90 " 25.09	1.25	37.50 " 37.69	1.88
12.50 " 12.69	.63	25.10 " 25.29	1.26	37.70 " 37.89	1.89

Estimation With Decimals

Sometimes an exact answer is not needed when you multiply or divide with decimals. At those times you should estimate the product or quotient. An estimate can be used to check if your answer is reasonable.

To estimate products of decimals, follow these steps:

Step 1   Round each decimal to its greatest whole number place-value position.

Step 2   Multiply.

Example

Step 1
$$\begin{array}{r} 1\,7\,.\,9\,2 \\ \times\quad 6\,.\,9 \\ \hline \end{array}$$
Round 17.92 to 20.
Round 6.9 to 7.

Step 2
$$\begin{array}{r} 2\,0 \\ \times\quad 7 \\ \hline 1\,4\,0 \end{array}$$
The product of 17.92 and 6.9 is about 140.

Practice

Estimate the products.

1. 59.8×2.08

2. 8.13×8.89

3. 241.4×29.68

_____   _____   _____

4.
$$\begin{array}{r} 7\,3\,3\,.\,9\,4 \\ \times\quad 2\,7\,.\,7 \\ \hline \end{array}$$

5.
$$\begin{array}{r} 0\,.\,9\,1\,5 \\ \times\quad 8\,2 \\ \hline \end{array}$$

6.
$$\begin{array}{r} 7\,8\,0 \\ \times\quad 5\,.\,8 \\ \hline \end{array}$$

_____   _____   _____

To estimate quotients of decimals, follow these steps:

Step 1   Round the divisor to the greatest whole number place-value position.

Step 2   Round the dividend so it can be divided evenly by the divisor.

Step 3   Divide.

Examples

A. $24.7 \div 4.7$

Step 1   Round 4.7 to 5.

Step 2   Round 24.7 to 25 because 25 can be divided evenly by 5.

Step 3
$$\begin{array}{r} 5 \\ 5\overline{)25} \end{array}$$

So, $24.7 \div 4.7$ is about 5.

B. $26.5\overline{)8472.1}$

Step 1   Round 26.5 to 30.

Step 2   Round 8472.1 to 9000 because 9000 can be divided evenly by 30.

Step 3
$$\begin{array}{r} 300 \\ 30\overline{)9000} \end{array}$$

So, $8472.1 \div 26.5$ is about 300.

Practice

Estimate the quotients.

7. $14.9 \div 3.2$

8. $5.5 \div 2.85$

9. $18.48 \div 6.09$

10. $5.6\overline{)12893}$

11. $5.6\overline{)128.93}$

12. $23.1\overline{)654.32}$

Problem Solving—Multiplying and Dividing Decimals

You have been solving word problems in the previous lessons. You should remember the following steps:

Step 1   Read the problem and underline the key words. These words will generally relate to some mathematics reasoning computation.

Step 2   Make a plan to solve the problem. Ask yourself, Should I add, subtract, multiply, divide, round, or compare? You may have to do more than one of these operations for the same problem. You may also be able to estimate your answer.

Step 3   Find the solution.

Step 4   Check the answer. Ask yourself, Is this answer reasonable? Did you find what you were asked for?

Here are some key words for multiplication and division.

Multiplication	Division
product	quotient
times	how many
of	average
apiece	shared
multiplied by	

Example

A. Three packages weigh 25.5, 17.3, and 45.25 grams, respectively. What is the average weight of the three? Round to the nearest hundredth.

Step 1   Determine the average weight of the three packages. The key word is **average.**

Step 2   The key word indicates which operation should occur—division. Before you can divide, you must first add to find the total gram weight of the three packages.

$$25.5 \text{ grams}$$
$$17.3 \text{ grams}$$
$$+45.25 \text{ grams}$$
$$\overline{88.05 \text{ grams}}$$

82

Step 3   Find the solution.

The average weight of the package is 29.35 grams.

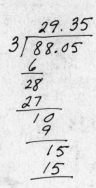

Step 4   Check the answer. Does it make sense that these three
packages ranging in weight from 17.3 to 45.25 could have
an average weight of 29.35? Yes, it is reasonable.

B. Constance and three friends gave $2.65 apiece to buy a gift for a
classmate. How much money did they have for the gift?

Step 1   Determine the amount of money collected for the gift. The
key word is **apiece.**

Step 2   The key word indicates which operation should
occur—multiplication.

Step 3   Find the solution.

$$\begin{array}{r} \$\ 2.65 \\ \times\ \ \ \ \ 4 \\ \hline \$10.60 \end{array}\ \text{amount collected}$$

Step 4   Check the answer. Does it make sense that $10.60 is the
total amount collected? Yes, the answer is reasonable.

Practice

Problem Solving

Solve the following, using the steps to solve word problems.

1. Estimate the cost of 609 kilowatt-hours
    of electricity at 7.9¢ a kilowatt.

2. The total length of a room divider is
    112.5 inches long. If each section is
    12.5 inches wide. How many sections
    are there in the divider?

Circle the correct answer.

3. The A team's broad jumpers recorded jumps of 24.26 meters, 21.5 meters, and 20.95 meters. The B team's jumpers jumped 26.42 meters, 18.75 meters, and 21.74 meters. What was the difference between the totals of the two teams?

   (1) .2 meters   (2) .24 meters
   (3) 2.16 meters   (4) 2.2 meters
   (5) none of these

4. A restaurant waitress earns $4.35 an hour. When she serves as a banquet waitress, she earns 55 cents more per hour. If she works 35 hours in the restaurant and another 12 hours on banquets, how much will she earn?

   (1) $152.25   (2) $171.50
   (3) $211.05   (4) $230.30
   (5) none of these

5. Seth rides the exercise bike at the health club. He started riding 15.25 minutes three times a week. He now has increased his riding time to 45.5 minutes per session. If he attends three times a week, how many more minutes per week is he now riding than when he began?

   (1) 45.75 minutes   (2) 125.5 minutes
   (3) 181.75 minutes   (4) 227.5 minutes
   (5) none of these

Problems 6 and 7 are related.

6. Moran's Landscaping Service needed 315.5 square feet of plywood, 80.8 feet of two-by-fours, and 32.4 feet of four-by-fours to build a storage shed. The plywood cost $3.75 per square foot, the two-by-fours cost $.45 per foot, and the four-by-fours cost $1.05 per foot. What will the lumber cost for this project? Round to nearest dollar.

   (1) $70   (2) $1183
   (3) $1200   (4) $1254
   (5) none of these

7. The roofing shingles for the shed cost $1390. If the estimate on the shed was $1250, how much is the project over the budget?

   (1) $110   (2) $120
   (3) $130   (4) $140
   (5) none of these

Multiplying and Dividing Decimals Using a Calculator

You can use your calculator to multiply or divide decimals.

Examples

A. To multiply .526 × .34, follow these steps:

Step 1   Press | C |. This will clear out the last problem.

Step 2   Press | · | | 5 | | 2 | | 6 |.

Step 3   Press | × |.

Step 4   Press | · | | 3 | | 4 |.

Step 5   Press | = |. Read the answer and write it down.   *.17884*

Step 6   Press | C | before starting the next problem.

B. To divide 25.01 by 6.3, follow these steps:

Step 1   Press | C |.

Step 2   Press | 2 | | 5 | | · | | 0 | | 1 |.

Step 3   Press | ÷ |.

Step 4   Press | 6 | | · | | 3 |.

Step 5   Press | = |.

Read the answer. Your calculator probably shows 3.9698412. Round the answer to the desired number of places. Write the answer.

Step 6   Press | C | before starting the next problem.

Practice

Multiply or divide. Use your calculator. Round your answer to the nearest hundredth.

1.   . 6 9 4
      × . 0 3

2. .7853 ÷ .08

3.   6 3 . 2
      × . 4 9 6

Find the products for the following.

1. $3.05 \times .004$

2.
$$\begin{array}{r} .44 \\ \times\ 14 \\ \hline \end{array}$$

3. $62.8 \times .0009$

4.
$$\begin{array}{r} 1.34 \\ \times\ 3.05 \\ \hline \end{array}$$

5. $.02715 \times 100$

6. $.02715 \times 1000$

Find the quotients for the following.

7. $0.9\overline{)0.0054}$

8. $2.04\overline{)22.644}$

9. $1.6\overline{)26.112}$

Round to the nearest decimal place named.

10. 51.85 ÷ 2.1
    (tenth)

    _____

11. .0552 ÷ .19
    (hundredth)

    _____

12. 2.715 ÷ 10
    (thousandths)

    _____

Estimate.

13. 2000 ÷ 4.72

    _____

14. 2 × 9.8

    _____

15. 3.3 × 0.85

    _____

Solve the following problems.

16. A small dress shop paid $1,567 for 100 sweaters. What was the cost per sweater?

    _____

17. A gas station pumped 675.25 gallons of gas in 2.5 hours. How much gas was pumped in an hour?

    _____

Use a calculator to solve.

18. There are forty-five 8-inch by 12-inch tiles in a box. Each tile cost $.69. If you buy two boxes of tiles, how much will you spend?

    _____

19. Toni has a Christmas saving plan. She saves $10.25 a week for 50 weeks. If she completes her plan, how much will she save?

    _____

Practice for Mastery of Decimals

Fill in the blank with the letter of the matching number.

1. _____ one and six tenths   a. .133
2. _____ five dollars and three cents   b. 5.03
3. _____ five and three hundredths   c. 1.6
4. _____ one hundred thirty-three thousandths   d. .000005
5. _____ five millionths   e. $5.03

Compare the following, using >, <, or =.

6. 8.09 _____ 8.90   7. .016 _____ .216

8. .00716 _____ .007160   9. .6 _____ .325

10. 5.6 ÷ 2 _____ 2.83   11. 3.57 ÷ 14 _____ 14.57

Write the decimals in order, from the smallest to the largest.

12. 35.1   3.5   .035   .35   _____ _____ _____ _____

13. .030   .035   .003   3.33   _____ _____ _____ _____

14. .088   .08   .8   .808   _____ _____ _____ _____

Add.

15. 23.4 + 2.031 + .984   16.   3 8 . 1 0 7   17. 1,315.93 + 7.6 + 0.95
                                          7 . 4
    _____   + _____. 5 2 9 1   _____

Multiply.

18. 1.357 × 10   19. 13.75 × 100   20. .1375 × 1,000

    _____   _____   _____

88

Subtract.

21. $1,260.3 - 11.345$

22.
$$761.21 - 70.003$$

23. $16.351 - 3.824$

Multiply.

24.
$$\$137.70 \times 12$$

25.
$$7.03 \times .05$$

26.
$$.03 \times .01$$

Divide.

27. $8\overline{)26.112}$

28. $.006\overline{).00618}$

29. $6.85\overline{)12.8095}$

30. $980.06 \div 10$

31. $980.06 \div 100$

32. $980.06 \div 1000$

33. Divide. Round the answer to the nearest tenth.

$.306\overline{)24.73}$

34. Divide. Round the answer to the nearest hundredth.

$4.1\overline{)7.387}$

Solve the following problems. Circle the correct answer.

35. 28.34 divided by .06, rounded to the nearest hundredth, is

    (1) 47.23   **(2)** 47.33

    (3) 472.32   **(4)** 473.33

    (5) 472.33

36. 136.32 multiplied by 24, rounded to the nearest tenth, is

    (1) 32.7   **(2)** 327.1

    (3) 3271.6   **(4)** 3271.7

    (5) 3.2716

Problems 37 and 38 are related.

37. The Hogan family ordered 35 square yards of carpeting. Fifteen yards were priced at $8.95 a square yard. The balance was $10.95 a square yard. How much did the carpet cost?

    (1) $134.25   **(2)** $219.00

    (3) $313.25   **(4)** $353.25

    (5) $325.13

38. If the Hogan family makes a downpayment of $53.25 and pays the balance in three equal payments, how much will each payment be?

    (1) $53.25   **(2)** $100.00

    (3) $125.00   **(4)** $150.00

    (5) $153.25

39. The electric company used kilowatt-hours to measure the amount of electricity usage. The Nelson family used a total of 80 kilowatt-hours during the month of January. In June, they used 165 kilowatt-hours. What was the difference in the Nelson's electric payments if each kilowatt-hour cost $.375? Round to the nearest cent.

    (1) $30.00   **(2)** $31.88

    (3) $61.87   **(4)** $61.88

    (5) $68.87

40. The weekend rental rate for a midsize car is $45.45 without insurance. The first hundred miles are free; every mile thereafter is $.12. If insurance is $5.95, and you travelled 495 miles that weekend, what was the total rental expense? Include insurance.

    (1) $51.45   **(2)** $53.35

    (3) $98.80   **(4)** $104.80

    (5) $15.95

41. The Classic Auto Body and Fender Shop estimates the time it takes to repair and to service cars. An oil change takes .25 of an hour, tire rotation takes .75 of an hour, and air conditioning service takes 1.25 hours. If the hourly labor rate is $10, what is the labor cost for all three services?

    (1) $10   **(2)** $17.50

    (3) $22.50   **(4)** $25.00

    (5) $25.20

42. A five-day vacation package was offered for $445.95. Lodging for each day was $45.25 and meals were $37.75 each day. How much was allowed for admissions to sightseeing activities?

    (1) $30.95   **(2)** $188.75

    (3) $226.25   **(4)** $415.00

    (5) $225.26

Decimals Posttest

Write the number value of each expression.

1. eight and four tenths _____

2. forty-eight millionths _____

3. eight hundred forty _____

4. forty-eight million _____

Compare the following, using >, <, or =.

5. (5×8) _____ $(6 \times 8) \div 2$

6. $2 \times (7 - 5)$ _____ $20 - 4$

7. $15 - (8 + 3)$ _____ $27 - (9 + 5)$

8. 0.007 _____ 0.0070

Write the following in order from largest to smallest.

9. 65 6.5 .065 65.1   _____ _____ _____ _____

10. .0079 730.0 07.3 73   _____ _____ _____ _____

Using the number line below, estimate the locations of the points in problems 11 through 14.

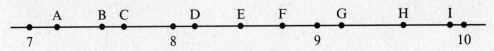

11. 9.9 _____

12. 7.6 _____

13. 8.7 _____

14. 7.2 _____

Find the sums.

15. $407.3 + 96.19 + 0.5$

16. $144 + 63 + 983 + 200$

17. $0.9326 + 99.3 + 7.25$

_____

_____

_____

Find the differences.

18.
```
  6,004
-   519
```

19. $\$43 - \$36.15$

20.
```
  649,529
- 229,799
```

_____

91

Find the products.

21.
$$\begin{array}{r} 1,565 \\ \times\ \ \ \ 202 \\ \hline \end{array}$$

22.
$$\begin{array}{r} 15.65 \\ \times\ \ \ 20.2 \\ \hline \end{array}$$

23. 1.25×10

_____

Find the quotients.

24. $5\overline{)85}$

25. $0.56\overline{)0.728}$

26. $5.2\overline{)0.6708}$

27. Round the quotient to the nearest tenth.

$5\overline{)68.38}$

28. Round the quotient to the nearest hundredth.

$3.7\overline{)244.507}$

29. Round the quotient to the nearest thousandth.

$38.7\overline{)132.338}$

Solve the following problems.

Problems 30 and 31 are related.

30. The Levin family of six paid $32 for admission to the races. Three ordered hot dogs at $1.25 each. Five had candy bars at 80 cents each, and all six had peanuts at 95 cents each. How much did they spend at the races?

(1) $25.25   (2) $35.75

(3) $39.75   (4) $45.45

(5) none of these

31. If Mr. Levin had $50 in his pocket before taking his family to the races, how much money did he have after the outing?

(1) $4.55   (2) $10.25

(3) $14.25   (4) $18.00

(5) $18.55

32. Four years ago a worker earned $5.25 an hour. Today, that worker earns $7.75 an hour. Find what she used to earn and what she now earns in a 40-hour week. What is the difference in weekly salary?

(1) $100   (2) $110

(3) $210   (4) $310

(5) $130

33. How far can a car travel in 14.25 hours if it averages 48 miles per hour in the first 7 hours and 52 miles per hour during the remaining time?

(1) 336 miles   (2) 377 miles

(3) 684 miles   (4) 713 miles

(5) 736 miles

Write fractions for the shaded portions.

1. 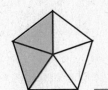 _____

2. _____

Circle the set of equivalent fractions.

3. $\frac{2}{3} = \frac{9}{12}$

4. $\frac{8}{10} = \frac{16}{20}$

Reduce the fractions to lowest terms.

5. $\frac{9}{24} =$ _____

6. $\frac{7}{14} =$ _____

Find the least common denominator for each of the following sets.

7. $\frac{7}{8}$ $\frac{7}{9}$ _____

8. $\frac{1}{8}$ $\frac{1}{2}$ _____

9. $\frac{1}{6}$ $\frac{1}{3}$ $\frac{5}{12}$ _____

10. $\frac{1}{5}$ $\frac{2}{3}$ $\frac{3}{10}$ _____

Compare the following fractions. Use >, <, or =.

11. $\frac{8}{32}$ _____ $\frac{1}{8}$

12. $\frac{5}{10}$ _____ $\frac{2}{4}$

13. $\frac{12}{16}$ _____ $\frac{2}{3}$

14. $\frac{10}{100}$ _____ $\frac{8}{80}$

Find the value of the following expressions.

15. $\frac{8 \cdot 2}{4} + \frac{7(4)}{2} =$

16. $\frac{28 + 5}{3(2) + 5} =$

17. $\frac{80}{4(5 - 3)} =$

18. $4(3 + 2) + 9 =$

_____ _____ _____ _____

93

Order of Operations

There are many ways to show multiplication. You can use the times symbol, 3×4; a raised dot, $3 \cdot 4$; or parentheses, (3) (4).

You can show division with the division symbol, $12 \div 3$, or by using a bar with the dividend on top and the divisor on the bottom, $\frac{12}{3}$.

Sometimes you will solve problems using a mixture of operations, including addition, subtraction, multiplication, and division. To do this, follow the order of operations shown below.

Step 1   Calculate everything inside parentheses.

Step 2   Simplify all exponents.

Step 3   Calculate multiplication or division operations, moving from left to right across the problem.

Step 4   Calculate additions or subtraction operations, moving from left to right across the problem.

Examples

A. Find the value of $3(2) + 8(4 + 1)$.

$3(2) + 8(4 + 1)$	Add inside parentheses.
$3(2) + 8(5)$	Multiply.
$6 + 40$	Add.
46	

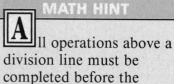

MATH HINT

All operations above a division line must be completed before the division can be performed.

B. Find the value of $\frac{16 - 4}{3(2)} + (4 \times 8)$.

$\frac{16 - 4}{3(2)} + (4 \times 8)$	To divide, the operations above the line must be completed first.
$\frac{12}{3(2)} + (4 \times 8)$	Multiply inside parentheses.
$\frac{12}{3(2)} + 32$	Multiply, then divide.
$2 + 32$	Add.
34	

Find the value of each expression. The first expression has been completed for you. Show your work.

1. $12 + 7 - 3(2) =$

    $19 \cdot - 6 =$ _____ 13 _____

2. $\dfrac{6(9 + 1)}{3} - 11 =$ _____

3. $\dfrac{2(8) - 4}{3} =$ _____

4. $5(14 + 7) - 12 =$ _____

5. $2 + 3(5) + 1 =$ _____

6. $3 \cdot 2 + 8(4 + 1) =$ _____

7. $14 + 2(30) =$ _____

8. $9(2)(4) - 22 =$ _____

9. $80 - 8(13 - 9) =$ _____

10. $6(2) + 5(2) + 6 =$ _____

11. $3(7) + \dfrac{15 - 3}{2} =$ _____

12. $\dfrac{72}{9} - \dfrac{21 + 3}{4} =$ _____

13. $\dfrac{9(4)}{3 \cdot 2} - \dfrac{8(6)}{4 \cdot 2} =$ _____

14. $5(2)(7) + 4(3) - 6 =$ _____

15. $2(6 + 3) - (7 + 1) =$ _____

16. $6(7) - 5 \cdot 4 =$ _____

Compare the following expressions, using >, <, or =.

17. $6(9) - 2(7)$ _____ $9(7) - 6(3)$

18. $27 - 2(4 - 1)$ _____ $5(4) - (7 + 2)$

19. $\dfrac{8 \cdot 2}{4}$ _____ $\dfrac{6(3 + 2)}{10}$

20. $\dfrac{4(7 + 2)}{3 + 9}$ _____ $\dfrac{(19 - 5)6}{4 + 2}$

21. 17 _____ $\dfrac{2(5 + 3)}{4}$

22. $3 + 3(6)$ _____ $3(7)$

23. $2(8 + 5)$ _____ $(6 + 4)3$

24. $5 \cdot 2 \cdot 7$ _____ $\dfrac{21 + 3}{4}$

Writing Fractions

A **fraction** is part of a whole unit. The numbers 1, 2, 3, and 4 tell how many whole units there are. Numbers like $\frac{1}{2}$, $\frac{2}{3}$, and $\frac{3}{4}$ are fractions.

Every fraction has a numerator and a denominator. The **denominator** tells how many equal parts make the whole. The **numerator** tells how many parts are being used. A short line separates the two numbers.

numerator $\frac{3}{4}$ three
denominator   fourths

When the numerator is greater than or equal to the denominator, as in $\frac{4}{2}$, you must divide. The value of $\frac{4}{2}$ is 2 whole units. When the numerator is less than the denominator, you will have a value less than one whole unit.

Examples

Fractions can tell what parts of these circles are shaded.

Fractions can measure lengths that are less than one whole inch.

A.

$\frac{5}{6}$

B.
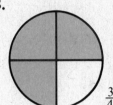
$\frac{3}{4}$

C.
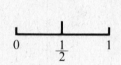

D.
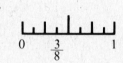

In addition to telling the number of shaded parts of an object or measures less than one inch, fractions tell how many equal parts that the whole has been divided. Study the examples below.

E.

2 equal parts
Halves

F.

3 equal parts
Thirds

G.

4 equal parts
Fourths

MATH HINT

After fourths, the names for fractional parts are like the counting numbers with **th** at the end of them.

Tell how many equal parts are in each figure. Then name the parts.

1.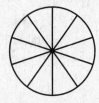

_____

2.

_____

3.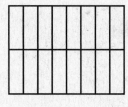

_____

4.

_____
_____

5.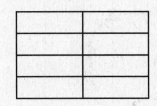

_____

6.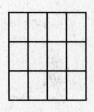

_____

Write a fraction to show what part or parts of the objects are shaded.

7.

_____

8.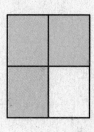

_____

9.

_____

10.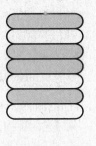

_____

11.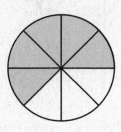

_____

12.

_____

Fractions can also tell about part of a group. The **denominator** tells
how many objects are in the group. The **numerator** tells how many
objects in the group are being talked about.

There are six flowers. Five of the flowers are
dark. What fraction are dark? $\frac{5}{6}$

Problem Solving

Solve the following problems. Write your answers as fractions.

13. There are _____ glasses.
    _____ glasses are full.
    What fraction are full? _____

14. There are _____ places for eggs.
    _____ eggs are in the carton.
    What fraction of places are
    filled? _____

15. Rhea had 5 inches of tubing. Write this as a fraction of a yard.
    (It takes 36 inches to make a yard.)   _____

16. Willis completed his homework in 50 minutes. Express this as a
    fraction of an hour.   _____

17. Macey works 8 hours a day. Write this as a fraction of the
    24-hour day.   _____

18. Susan earns $320 a month. She saves $70 a month. What
    fraction of the amount she earns does she save?   _____

19. A year has twelve months. Each season has three months. What
    fraction of the year is the summer season?   _____

Writing Equal Fractions

---- **Example** ----

The boxes to the right are equal in area. The box on the left has been divided into 6 equal parts. Three parts are shaded. The box on the right has been divided into 2 equal parts. One of those is shaded. The shaded portions are equal. Thus, $\frac{3}{6}$ and $\frac{1}{2}$ are equal.

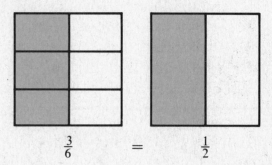

$$\frac{3}{6} = \frac{1}{2}$$

---- **Practice** ----

Give the equal fractions for shaded portions. Write in the number.

1.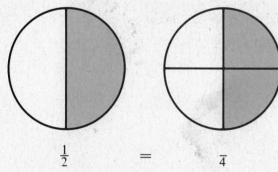

$$\frac{1}{2} = \frac{}{4}$$

2.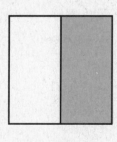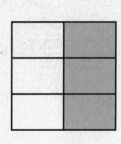

$$\frac{}{2} = \frac{}{6}$$

3.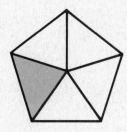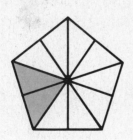

$$\frac{}{5} = \frac{}{10}$$

4.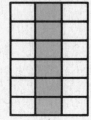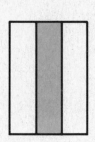

$$\frac{}{18} = \frac{}{3}$$

5.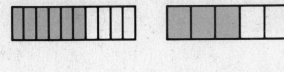

$$\frac{}{10} = \frac{}{5}$$

6.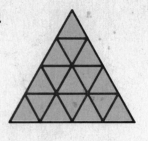

$$\frac{}{16} = \frac{}{1}$$

To write equal fractions, start with one fraction. Then either

1. multiply both the numerator and the denominator by the same number. (Never use zero.) This raises the fraction to higher terms. **or** **2.** divide both the numerator and the denominator by the same number. (Never use zero.) This reduces the fraction to lower terms.

--- **Examples** ---

A. Write fractions equal to $\frac{1}{3}$.

$$\overset{1 \times 2 \qquad 1 \times 3 \qquad 1 \times 4 \qquad 1 \times 5}{\frac{1}{3} \; = \; \frac{2}{6} \; = \; \frac{3}{9} \; = \; \frac{4}{12} \; = \; \frac{5}{15}}$$
$$3 \times 2 \qquad 3 \times 3 \qquad 3 \times 4 \qquad 3 \times 5$$

B. Write fractions equal to $\frac{12}{24}$.

$$\overset{12 \div 2 \qquad 12 \div 4 \qquad 12 \div 12}{\frac{12}{24} \; = \; \frac{6}{12} \; = \; \frac{3}{6} \; = \; \frac{1}{2}}$$
$$24 \div 2 \qquad 24 \div 4 \qquad 24 \div 12$$

--- **Practice** ---

Find the missing numerator or denominator in the pairs of equal fractions below. The first two are done as examples.

7. $\frac{12}{27} = \frac{}{9}$

27 was divided by 3 to get 9.

Divide 12 by 3 and get 4.

$\frac{12}{27} = \frac{4}{9}$

8. $\frac{1}{15} = \frac{4}{}$

1 was multiplied by 4 to get 4.

Multiply 15 by 4 and get 60.

$\frac{1}{15} = \frac{4}{60}$

9. $\frac{32}{40} = \frac{}{5}$

10. $\frac{2}{5} = \frac{}{100}$

11. $\frac{20}{35} = \frac{}{7}$

12. $\frac{12}{36} = \frac{}{3}$

13. $\frac{3}{5} = \frac{}{25}$

14. $\frac{48}{56} = \frac{6}{}$

15. $\frac{3}{4} = \frac{24}{}$

16. $\frac{25}{60} = \frac{}{12}$

17. $\frac{35}{63} = \frac{5}{}$

18. $\frac{7}{15} = \frac{14}{}$

19. $\frac{1}{9} = \frac{}{54}$

20. $\frac{9}{20} = \frac{18}{}$

Fractions of an inch

Inches can be divided into equal parts. Inches are divided into halves, then the halves are divided into fourths, and so on.

Here are the fractions of an inch. Study the equal parts.

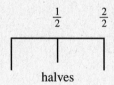
halves

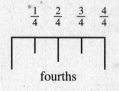
fourths

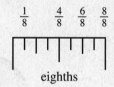
eighths

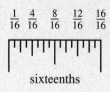
sixteenths

You can find many equal fractions using the fractions of an inch.

Use the figures above to find the numerators.

21. $\frac{1}{2} = \frac{}{4}$

22. $\frac{1}{2} = \frac{}{8}$

23. $\frac{1}{2} = \frac{}{16}$

24. $\frac{1}{4} = \frac{}{8}$

25. $\frac{1}{4} = \frac{}{16}$

26. $\frac{1}{8} = \frac{}{16}$

27. $\frac{3}{4} = \frac{}{8}$

28. $\frac{3}{4} = \frac{}{16}$

Are these fractions equal? Answer *yes* or *no*.

29. $\frac{2}{2}$ $\frac{8}{8}$ _____

30. $\frac{4}{4}$ $\frac{16}{16}$ _____

31. $\frac{16}{16}$ $\frac{1}{4}$ _____

32. $\frac{1}{8}$ $\frac{1}{2}$ _____

33. $\frac{3}{4}$ $\frac{3}{16}$ _____

34. $\frac{3}{8}$ $\frac{6}{16}$ _____

Do these equal one inch? Answer *yes* or *no*.

35. $\frac{2}{2}$ inch _____

36. $\frac{8}{8}$ inch _____

Reducing Fractions to Lowest Terms

To reduce a fraction to its lowest terms, **divide** both the **numerator** and the **denominator** by the **same number.** Remember, never divide by zero. Continue to divide until the only number that will divide both the numerator and the denominator is **one.**

Examples

A. Reduce $\frac{4}{8}$ to lowest terms.

First, try dividing both the numerator and the denominator by the numerator.

$$\frac{4}{8} = \frac{4 \div 4}{8 \div 4} = \frac{1}{2}$$

$\frac{1}{2}$ is the answer.

B. Reduce $\frac{16}{36}$ to lowest terms.

The numerator does not divide into the denominator. Divide by another number.

$$\frac{16}{36} = \frac{16 \div 2}{36 \div 2} = \frac{8}{18}$$ Both 8 and 18 can be divided by 2.

$$\frac{8}{18} = \frac{8 \div 2}{18 \div 2} = \frac{4}{9}$$

$\frac{4}{9}$ is the answer.

Practice

Reduce these fractions to lowest terms.

1. $\frac{5}{15} =$ _____

2. $\frac{2}{12} =$ _____

3. $\frac{10}{14} =$ _____

4. $\frac{8}{12} =$ _____

5. $\frac{3}{24} =$ _____

6. $\frac{16}{18} =$ _____

7. $\frac{9}{15} =$ _____

8. $\frac{40}{50} =$ _____

9. $\frac{9}{30} =$ _____

10. $\frac{8}{50} =$ _____

11. $\frac{12}{16} =$ _____

12. $\frac{17}{34} =$ _____

13. $\frac{28}{36} =$ _____

14. $\frac{6}{27} =$ _____

15. $\frac{48}{56} =$ _____

16. $\frac{12}{27} =$ _____

17. $\frac{9}{12} =$ _____

18. $\frac{14}{35} =$ _____

19. $\frac{9}{27} =$ _____

20. $\frac{21}{27} =$ _____

21. $\frac{20}{26} =$ _____

22. $\frac{28}{32} =$ _____

23. $\frac{8}{32} =$ _____

24. $\frac{24}{36} =$ _____

25. $\frac{15}{60} =$ _____

26. $\frac{20}{50} =$ _____

27. $\frac{6}{54} =$ _____

28. $\frac{48}{72} =$ _____

29. $\frac{18}{63}$ = _____   **30.** $\frac{12}{40}$ = _____   **31.** $\frac{14}{16}$ = _____   **32.** $\frac{48}{96}$ = _____

33. $\frac{50}{100}$ = _____   **34.** $\frac{44}{55}$ = _____   **35.** $\frac{24}{54}$ = _____   **36.** $\frac{3}{24}$ = _____

37. $\frac{15}{24}$ = _____   **38.** $\frac{20}{24}$ = _____   **39.** $\frac{25}{35}$ = _____   **40.** $\frac{5}{45}$ = _____

41. $\frac{12}{32}$ = _____   **42.** $\frac{50}{75}$ = _____   **43.** $\frac{10}{15}$ = _____   **44.** $\frac{11}{99}$ = _____

45. $\frac{14}{21}$ = _____   **46.** $\frac{25}{55}$ = _____   **47.** $\frac{25}{125}$ = _____   **48.** $\frac{8}{14}$ = _____

49. $\frac{7}{42}$ = _____   **50.** $\frac{32}{96}$ = _____   **51.** $\frac{9}{54}$ = _____   **52.** $\frac{10}{12}$ = _____

53. $\frac{16}{48}$ = _____   **54.** $\frac{20}{22}$ = _____   **55.** $\frac{80}{100}$ = _____   **56.** $\frac{6}{9}$ = _____

Problem Solving

Solve the following problems. Reduce to lowest terms. Circle the correct answer.

57. Kay's Shoe Store had 125 pairs of shoes on sale. After selling some, there were 75 pairs left. What fractional part of the pairs were sold?

   (1) $\frac{50}{125}$ **(2)** $\frac{2}{5}$

   (3) $\frac{75}{125}$ **(4)** $\frac{3}{5}$

   (5) none of these

58. Milano earns $160 a week and saves $20. What fractional part of his earnings does he save?

   (1) $\frac{20}{160}$ **(2)** $\frac{1}{8}$

   (3) $\frac{140}{160}$ **(4)** $\frac{7}{8}$

   (5) none of these

59. Taylor exercises 15 minutes per hour. What fractional part of an hour does he exercise?

   (1) $\frac{1}{4}$ **(2)** $\frac{2}{5}$

   (3) $\frac{1}{2}$ **(4)** $\frac{3}{4}$

   (5) none of these

60. Melissa slept 9 hours. What fractional part of the day did she sleep?

   (1) $\frac{1}{12}$ **(2)** $\frac{3}{8}$

   (3) $\frac{1}{2}$ **(4)** $\frac{5}{8}$

   (5) none of these

61. Four roses in a dozen were yellow. The remainder were red. What fractional part of the roses were red?

   (1) $\frac{1}{3}$ **(2)** $\frac{1}{2}$

   (3) $\frac{2}{3}$ **(4)** $\frac{3}{4}$

   (5) none of these

Everyday Measurements

There are 60 minutes in an hour. Write the following as fractions of an hour.

1. $\frac{15}{60}$ is _____ of an hour.

2. $\frac{20}{60}$ is _____ of an hour.

3. $\frac{30}{60}$ is _____ of an hour.

4. $\frac{40}{60}$ is _____ of an hour.

5. $\frac{45}{60}$ is _____ of an hour.

There are 100 pennies in a dollar. Write the following as fractions of a dollar.

6. $\frac{20}{100}$ is _____ of a dollar.

7. $\frac{25}{100}$ is _____ of a dollar.

8. $\frac{50}{100}$ is _____ of a dollar.

9. $\frac{75}{100}$ is _____ of a dollar.

10. $\frac{80}{100}$ is _____ of a dollar.

There are 36 inches in a yard. Write the following as fractions of a yard.

11. $\frac{9}{36}$ is _____ of a yard.

12. $\frac{12}{36}$ is _____ of a yard.

13. $\frac{18}{36}$ is _____ of a yard.

14. $\frac{24}{36}$ is _____ of a yard.

15. $\frac{27}{36}$ is _____ of a yard.

Common Denominators

Denominators that are the same are called **common denominators.**
To find a common denominator for two or more fractions, find a
number that all the denominators divide into evenly. Follow **one** of
these steps:

Step 1   Choose the denominator that the others divide into evenly.

   or

Step 2   Multiply the denominators together to get a common
denominator.

Examples

Step 1 $\frac{1}{3}$ and $\frac{2}{9}$ 3 divides evenly
into 9.
Use 9 as the common
denominator.

Step 2 $\frac{3}{5}$ and $\frac{1}{4}$ $5 \times 4 = 20$
Use 20 as the common
denominator.

Practice

Use Step 1 to find the common denominator.

1. $\frac{1}{5}$ $\frac{3}{10}$ _____

2. $\frac{1}{2}$ $\frac{3}{8}$ _____

3. $\frac{3}{4}$ $\frac{5}{12}$ _____

4. $\frac{1}{5}$ $\frac{4}{15}$ _____

5. $\frac{2}{3}$ $\frac{3}{15}$ _____

6. $\frac{12}{17}$ $\frac{5}{34}$ _____

Use Step 2 to find the common denominator.

7. $\frac{1}{8}$ $\frac{2}{3}$ _____

8. $\frac{2}{11}$ $\frac{1}{3}$ _____

9. $\frac{5}{6}$ $\frac{1}{5}$ _____

10. $\frac{5}{8}$ $\frac{3}{10}$ _____

11. $\frac{1}{13}$ $\frac{1}{2}$ _____

12. $\frac{5}{7}$ $\frac{2}{5}$ _____

Finding Least Common Denominators

The smallest possible common denominator is called the **least common denominator.** To find the least common denominator, follow these steps.

Step 1   List multiples of each denominator. A **multiple** of a number is the number multiplied by 1, 2, 3, or 4, and so on.

Step 2   Look for the first number in a list that matches a number in the other list. This number is the **least common denominator.**

Examples

A. Find the least common denominator of $\frac{1}{4}$ and $\frac{7}{10}$.

Step 1 $\frac{1}{4}$ and $\frac{7}{10}$

Multiples of 4	Multiples of 10
$4 \times 1 = 4$	$10 \times 1 = 10$
$4 \times 2 = 8$	$10 \times 2 = 20$
$4 \times 3 = 12$	
$4 \times 4 = 16$	
$4 \times 5 = 20$	

Step 2   The first number that matches in each list is 20.
20 is the least common denominator. Notice that 20 is the smallest number that both 4 and 10 can divide into evenly.

B. Find the least common denominator of $\frac{2}{3}$ and $\frac{3}{4}$.

Step 1 $\frac{2}{3}$ and $\frac{3}{4}$

Multiples of 3	Multiples of 4
$3 \times 1 = 3$	$4 \times 1 = 4$
$3 \times 2 = 6$	$4 \times 2 = 8$
$3 \times 3 = 9$	$4 \times 3 = 12$
$3 \times 4 = 12$	

Step 2   The first number that matches in each list is 12.
12 is the **least common denominator.** Notice that 12 is the smallest number that 3 and 4 can divide into evenly.

Find the least common denominator for each set of numbers.

13. $\frac{4}{5}$ $\frac{5}{6}$ _____   **14.** $\frac{5}{8}$ $\frac{1}{6}$ _____   **15.** $\frac{7}{12}$ $\frac{1}{3}$ _____   **16.** $\frac{1}{2}$ $\frac{5}{6}$ _____

17. $\frac{4}{5}$ $\frac{1}{3}$ _____   **18.** $\frac{3}{4}$ $\frac{4}{5}$ _____   **19.** $\frac{5}{8}$ $\frac{1}{7}$ _____   **20.** $\frac{5}{10}$ $\frac{3}{8}$ _____

21. $\frac{5}{12}$ $\frac{11}{36}$ _____   **22.** $\frac{1}{9}$ $\frac{5}{27}$ _____   **23.** $\frac{7}{8}$ $\frac{5}{9}$ _____   **24.** $\frac{5}{13}$ $\frac{11}{39}$ _____

25. $\frac{5}{6}$ $\frac{5}{36}$ _____   **26.** $\frac{4}{9}$ $\frac{1}{5}$ _____   **27.** $\frac{12}{17}$ $\frac{5}{34}$ _____   **28.** $\frac{1}{2}$ $\frac{8}{11}$ _____

29. $\frac{7}{16}$ $\frac{5}{8}$ _____   **30.** $\frac{3}{8}$ $\frac{1}{3}$ _____   **31.** $\frac{1}{4}$ $\frac{6}{7}$ _____   **32.** $\frac{1}{3}$ $\frac{2}{5}$ _____

33. $\frac{1}{3}$ $\frac{3}{4}$ $\frac{1}{2}$ _____   **34.** $\frac{13}{20}$ $\frac{1}{5}$ $\frac{3}{4}$ _____   **35.** $\frac{11}{16}$ $\frac{1}{4}$ $\frac{3}{8}$ _____   **36.** $\frac{5}{9}$ $\frac{3}{4}$ $\frac{1}{3}$ _____

37. $\frac{1}{2}$ $\frac{4}{5}$ $\frac{5}{8}$ _____   **38.** $\frac{3}{4}$ $\frac{7}{10}$ $\frac{1}{5}$ _____   **39.** $\frac{5}{9}$ $\frac{1}{3}$ $\frac{11}{27}$ _____   **40.** $\frac{4}{7}$ $\frac{2}{3}$ $\frac{1}{2}$ _____

41. $\frac{1}{6}$ $\frac{3}{4}$ $\frac{1}{2}$ _____   **42.** $\frac{1}{3}$ $\frac{1}{2}$ $\frac{1}{10}$ _____   **43.** $\frac{1}{3}$ $\frac{2}{7}$ $\frac{4}{21}$ _____   **44.** $\frac{5}{16}$ $\frac{3}{4}$ $\frac{5}{8}$ _____

45. $\frac{1}{2}$ $\frac{2}{5}$ $\frac{7}{10}$ _____   **46.** $\frac{5}{8}$ $\frac{3}{4}$ $\frac{2}{3}$ _____   **47.** $\frac{3}{4}$ $\frac{1}{3}$ $\frac{1}{6}$ _____   **48.** $\frac{1}{3}$ $\frac{5}{6}$ $\frac{3}{5}$ _____

49. $\frac{5}{7}$ $\frac{1}{2}$ $\frac{9}{14}$ _____   **50.** $\frac{5}{8}$ $\frac{1}{2}$ $\frac{1}{4}$ _____   **51.** $\frac{1}{16}$ $\frac{11}{12}$ $\frac{7}{8}$ _____   **52.** $\frac{7}{9}$ $\frac{3}{4}$ _____

Comparing Fractions

To compare fractions, follow these steps:

Step 1   Write all fractions with a common denominator.

Step 2   Compare the numerators.

Examples

A. Compare $\frac{5}{8}$ and $\frac{3}{8}$. Use >, <, or =.
The denominators are already the same.
Compare the numerators.

$$5 > 3$$

So $\frac{5}{8} > \frac{3}{8}$.

B. Compare $\frac{1}{3}$ and $\frac{2}{5}$. Use >, <, or =.
A common denominator of these
fractions is 15.

$$\frac{1}{3} = \frac{}{15} \qquad \frac{1 \times 5}{3 \times 5} = \frac{5}{15}$$

$$\frac{2}{5} = \frac{}{15} \qquad \frac{2 \times 3}{5 \times 3} = \frac{6}{15}$$

Compare the numerators.

$$5 < 6$$

So $\frac{5}{15} < \frac{6}{15}$.

$$\frac{1}{3} < \frac{2}{5}$$

Practice

Compare the following fractions. Use >, <, or =.

1. $\frac{7}{8}$ ——— $\frac{3}{8}$

2. $\frac{3}{4}$ ——— $\frac{75}{100}$

3. $\frac{4}{50}$ ——— $\frac{5}{25}$

4. $\frac{8}{9}$ ——— $\frac{1}{7}$

5. $\frac{5}{11}$ ——— $\frac{8}{11}$

6. $\frac{3}{5}$ ——— $\frac{1}{3}$

7. $\frac{1}{5}$ ——— $\frac{1}{2}$

8. $\frac{18}{20}$ ——— $\frac{8}{10}$

9. $\frac{7}{8}$ ——— $\frac{2}{3}$

10. $\frac{1}{4}$ ——— $\frac{1}{3}$

11. $\frac{5}{12}$ ——— $\frac{1}{3}$

12. $\frac{5}{9}$ ——— $\frac{2}{3}$

13. $\frac{2}{11}$ ——— $\frac{1}{2}$

14. $\frac{3}{9}$ ——— $\frac{15}{45}$

15. $\frac{5}{6}$ ——— $\frac{3}{7}$

16. $\frac{1}{3}$ ——— $\frac{3}{4}$

17. $\frac{3}{8}$ ——— $\frac{9}{24}$

18. $\frac{3}{4}$ ——— $\frac{2}{7}$

19. $\frac{3}{16}$ —— $\frac{1}{8}$

20. $\frac{3}{5}$ —— $\frac{3}{8}$

21. $\frac{2}{3}$ —— $\frac{3}{4}$

22. $\frac{1}{2}$ —— $\frac{2}{7}$

23. $\frac{5}{20}$ —— $\frac{4}{5}$

24. $\frac{2}{25}$ —— $\frac{4}{50}$

Order the following fractions from the largest to smallest.

25. $\frac{3}{5}$ $\frac{2}{5}$ $\frac{1}{5}$

26. $\frac{5}{8}$ $\frac{6}{8}$ $\frac{3}{8}$

27. $\frac{7}{12}$ $\frac{11}{12}$ $\frac{10}{12}$

_____

_____

_____

28. $\frac{1}{2}$ $\frac{1}{6}$ $\frac{1}{4}$

29. $\frac{2}{15}$ $\frac{2}{5}$ $\frac{7}{15}$

30. $\frac{6}{15}$ $\frac{1}{3}$ $\frac{1}{5}$

_____

_____

_____

31. $\frac{2}{5}$ $\frac{4}{5}$ $\frac{11}{20}$

32. $\frac{1}{3}$ $\frac{4}{5}$ $\frac{2}{3}$ $\frac{11}{15}$

33. $\frac{1}{2}$ $\frac{5}{7}$ $\frac{4}{7}$ $\frac{13}{14}$

_____

_____

_____

34. $\frac{5}{6}$ $\frac{4}{5}$ $\frac{2}{3}$ $\frac{1}{2}$

35. $\frac{3}{7}$ $\frac{1}{4}$ $\frac{1}{2}$ $\frac{2}{7}$

36. $\frac{23}{28}$ $\frac{5}{7}$ $\frac{3}{4}$ $\frac{1}{2}$

_____

_____

_____

37. $\frac{3}{5}$ $\frac{3}{7}$ $\frac{2}{5}$ $\frac{17}{35}$

38. $\frac{1}{5}$ $\frac{1}{2}$ $\frac{2}{5}$ $\frac{9}{10}$

39. $\frac{6}{7}$ $\frac{5}{6}$ $\frac{21}{42}$ $\frac{5}{7}$

_____

_____

_____

Problem Solving—Fractions

The steps you have learned to solve word problems can be used with word problems that deal with fractions. Recall the following steps:

Step 1 Read the problem and underline the key words. These words will generally relate to some mathematics reasoning computation.

Step 2 Make a plan to solve the problem. Ask yourself, Should I add, subtract, multiply, divide, round, or compare? You may have to do more than one of these operations for the same problem.

Step 3 Find the solution. Use your math knowledge to find your answer.

Step 4 Check the answer. Ask yourself, Is the answer reasonable? Did you find what you were asked?

Key fraction words to remember are as follows:

Numerator tells how many parts or objects in the group are being talked about. It is the top number of the fraction.

Denominator tells how many parts or objects are in the group. It is the bottom number of the fraction.

Example

A. Jonathan bought a pound of meat from the butcher. He used 4 ounces for a recipe. What fraction of the amount he purchased is left?

Step 1 Determine how many ounces are left and express the answer in a fraction. The key words are **left** and **fraction**.

Step 2 The key word **left** indicates which operation should occur—subtraction. Before you can perform this operation, you must also determine how many ounces are in a pound. If you said 16, you are correct.

Step 3 Find the solution.

$$
\begin{array}{r}
16 \text{ ounces in a pound} \\
-\ 4 \text{ ounces used} \\
\hline
12 \text{ ounces left}
\end{array}
\qquad
\begin{array}{l}
\text{Fraction} \\
\frac{12}{16}
\end{array}
$$

Reduce $\frac{12}{16}$ to lowest terms. The answer is $\frac{3}{4}$.

Step 4 Check the answer. Does it make sense that 12 ounces was left and $\frac{12}{16}$, or $\frac{3}{4}$, is the fraction form? Yes, the answer is reasonable.

Practice

Solve the following problems.

1. Which fraction is larger, $\frac{1}{4}$ or $\frac{6}{7}$? _____

2. Which fraction is smallest, $\frac{1}{6}$, $\frac{1}{5}$, or $\frac{7}{30}$? _____

3. Troy said he read $\frac{5}{12}$ of a book, and Eduardo said he read $\frac{1}{3}$ of the same book. Which person read more? _____

4. Among the fractions $\frac{3}{4}$, $\frac{5}{6}$, and $\frac{2}{3}$, which is equal to $\frac{18}{24}$? _____

5. Between the fractions $\frac{1}{8}$ and $\frac{2}{3}$, which fraction is larger than $\frac{5}{24}$? _____

6. Cars A and B started a 30-mile trip at the same time and to the same place. At a given time, Car A traveled $\frac{15}{16}$ of the distance and Car B traveled $\frac{31}{32}$ of the distance. Which car was ahead at that given time? _____

7. Two screws were measured. Screw X measured $\frac{1}{4}$ of an inch. Screw Y measured $\frac{1}{2}$ inch. Which screw was longer? _____

8. Which of the following figures has the greatest portion shaded?

   A. [figure] B. [figure] C. [figure] _____

   $\frac{2}{3}$ $\frac{2}{4}$ $\frac{2}{8}$

9. Among the fractions $\frac{4}{15}$, $\frac{2}{5}$, and $\frac{7}{15}$, which is equal to $\frac{10}{25}$? _____

10. Is $\frac{1}{4}$ of a dozen greater than, less than, or equal to $\frac{3}{12}$ of a dozen? _____

11. There are four quarters in a football game, and three quarters have been played. Express as a fraction the unfinished portion of the game.

_____

12. There are 11 rows in a garden. Four have vegetables and the rest have melons. What fraction of the garden has melons?

_____

13. There are 10 pins that the bowler attempts to knock down. If 3 pins remain standing, express as a fraction the number of pins down.

_____

14. An ice tray has 12 sections. Four sections have cubes. What fraction of the tray has no cubes?

_____

15. There are 36 ice cubes in an ice bucket. Using the information from Problem 14, how many trays did it take to make those cubes? Express as a fraction.

_____

16. A banquet table seats 8 people. Three seats are empty. Express the number of seated people as a fraction.

_____

17. If 320 people purchased tickets to a banquet, how many tables of eight will be needed? Express as a fraction.

_____

18. There are fifty dimes in a coin purse. How many dollars does this represent? Express as a fraction.

_____

19. If one quarter of a dollar has been spent, express the unspent value as a fraction.

_____

20. The Pacific Ocean covers approximately $\frac{4}{12}$ of the earth's surface. Express the fraction in lowest terms.

_____

Posttest

Write fractions for the shaded portions of the figures. Reduce to lowest terms.

1. 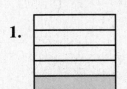 _____

2. _____

3. 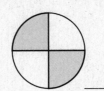 _____

4. _____

Circle the sets of equivalent fractions.

5. $\frac{5}{8} = \frac{25}{40}$ 6. $\frac{3}{4} = \frac{6}{28}$ 7. $\frac{18}{24} = \frac{3}{4}$ 8. $\frac{5}{10} = \frac{1}{4}$

Reduce the following fractions to lowest terms.

9. $\frac{16}{20} =$ _____   10. $\frac{25}{30} =$ _____   11. $\frac{8}{24} =$ _____   12. $\frac{11}{22} =$ _____

Find the least common denominator for each of the following sets.

13. $\frac{1}{3}$ $\frac{1}{12}$ $\frac{5}{6}$ _____   14. $\frac{5}{9}$ $\frac{1}{18}$ _____   15. $\frac{2}{3}$ $\frac{3}{10}$ _____   16. $\frac{7}{8}$ $\frac{5}{6}$ $\frac{1}{3}$ _____

Compare these fractions. Use >, <, or =.

17. $\frac{5}{7}$ _____ $\frac{7}{10}$ 18. $\frac{8}{9}$ _____ $\frac{9}{10}$ 19. $\frac{2}{3}$ _____ $\frac{10}{15}$ 20. $\frac{16}{25}$ _____ $\frac{4}{5}$

Find the value of the following expressions.

21. $\frac{4(7+3)}{5} - \frac{24}{12} =$ 22. $8(5) - 6(3) =$ 23. $35 - \frac{5+5}{7-2} =$

_____   _____   _____

Solve the problems. Write the fractions in lowest terms.

24. Casey used 3 eggs to make a cake. Write this as a fraction of a dozen.

_____

25. A car's gas tank holds 16 gallons. The tank has 12 gallons now. Write this as a fraction.

_____

Mixed Numbers

Write a mixed number for each picture.

1.   _____

2.   _____

Compare the following mixed numbers using >, <, or =.

3. $6\frac{16}{21}$ _____ $6\frac{2}{3}$ **4.** $11\frac{3}{8}$ _____ $11\frac{9}{24}$ **5.** $9\frac{5}{7}$ _____ $9\frac{29}{35}$ **6.** $2\frac{12}{20}$ _____ $2\frac{3}{5}$

Change the improper fractions to mixed or whole numbers.

7. $\frac{75}{10}$ _____   **8.** $\frac{16}{2}$ _____   **9.** $\frac{56}{5}$ _____   **10.** $\frac{51}{12}$ _____

Change the following mixed numbers to improper fractions.

11. $5\frac{1}{4}$ _____   **12.** $14\frac{1}{5}$ _____   **13.** $6\frac{5}{8}$ _____   **14.** $4\frac{4}{15}$ _____

Divide. Write the quotients as mixed numbers.

15. $5\overline{)256}$ **16.** $9\overline{)2,172}$ **17.** $63\overline{)7,215}$ **18.** $24\overline{)774}$

Writing Mixed Numbers

Kelley is a manager at Joe's Sandwich Shop. She cut 3 pies into 8 equal pieces. She served 5 pieces. At the end of the day, she had 2 whole pies and $\frac{3}{8}$ of a pie left.

$2\frac{3}{8}$ is called a mixed number.

A **mixed number** is a whole number plus a fraction.

Example

This is what Kelley had left.

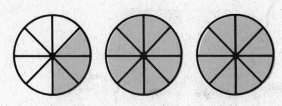

You write it as a mixed number: whole number → $2\frac{3}{8}$ ← fraction

---

Practice

Write mixed numbers for the shaded portion of these figures.

1. 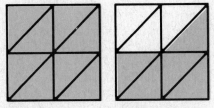 _____

2.  _____

3. 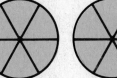 _____

4. 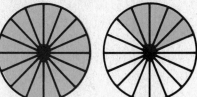 _____

5. 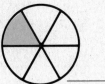 _____

Reading a Ruler

A. If each inch is divided into 4 spaces, each space is $\frac{1}{4}$ inch.

B. If each inch is divided into 8 spaces, each space is $\frac{1}{8}$ inch.

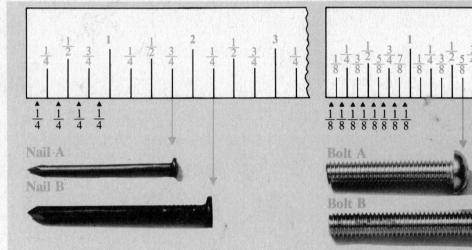

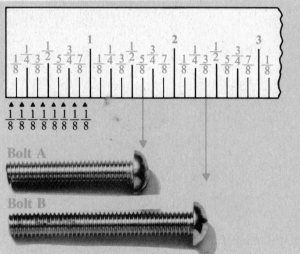

How long is nail A? $1\frac{3}{4}$ inches

How long is nail B? _____

How long is bolt A? $1\frac{5}{8}$ inches

How long is bolt B? _____

Most rulers are not labeled $\frac{1}{4}$, $\frac{1}{8}$, or $\frac{1}{16}$. Count the spaces in an inch to decide what fraction each mark stands for.

If each inch is divided into 16 spaces, each space is $\frac{1}{16}$ inch.

C.

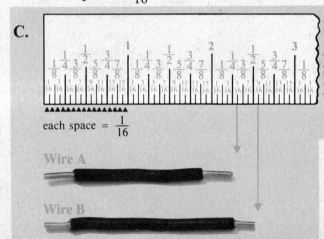

each space = $\frac{1}{16}$

Wire A

Wire B

Notice the lines are of different heights.

1 inch
$\frac{1}{2}$ inch
$\frac{1}{4}$ inch
$\frac{1}{8}$ inch
$\frac{1}{16}$ inch

How long is wire A? $2\frac{5}{16}$ inches.

How long is wire B? _____

Use a ruler to measure the length of the following items. The first is completed for you. Express the measurement as a mixed number.

1. $3\frac{5}{8}''$

2. _____

3. _____

4. _____

5. _____

6. _____

7. _____

8. _____

Comparing and Ordering Mixed Numbers

To compare mixed numbers using > or <, follow these steps:

Step 1   First compare the whole numbers.

Step 2   When the whole numbers are the same, compare the fractions. If the denominators of the fractions are the same, compare the numerators.

Step 3   If the denominators are different, find a common denominator. Then compare the numerators.

Examples

A. Compare $1\frac{1}{2}$ and $2\frac{3}{4}$.

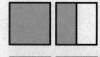

1. Compare the whole numbers.

$$1 < 2$$

So, $1\frac{1}{2} < 2\frac{3}{4}$.

B. Compare $5\frac{3}{5}$ and $5\frac{4}{5}$.

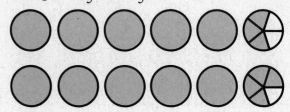

1. The whole numbers are the same.

$$5 = 5$$

2. Compare the fractions.

$$\frac{3}{5} < \frac{4}{5}$$

So, $5\frac{3}{5} < 5\frac{4}{5}$.

C. Compare $7\frac{1}{4}$ and $7\frac{1}{12}$.

1. The whole numbers are the same.

2. Compare the fractions. The denominators are different.

3. Find a common denominator and compare the fractions.

$$7 = 7$$

$$\frac{1}{4} \qquad \frac{1}{12}$$

$$\frac{3}{12} > \frac{1}{12}$$

So, $7\frac{1}{4} > 7\frac{1}{12}$.

Compare the mixed numbers using >, <, or =.

1. $5\frac{1}{3}$ _____ $5\frac{5}{15}$

2. $6\frac{3}{7}$ _____ $6\frac{8}{21}$

3. $2\frac{5}{7}$ _____ $2\frac{21}{28}$

4. $1\frac{2}{3}$ _____ $1\frac{4}{5}$

5. $1\frac{1}{2}$ _____ $1\frac{7}{10}$

6. $11\frac{7}{12}$ _____ $11\frac{1}{2}$

7. $4\frac{1}{2}$ _____ $4\frac{3}{8}$

8. $11\frac{25}{30}$ _____ $11\frac{30}{36}$

9. $3\frac{1}{2}$ _____ $3\frac{1}{3}$

10. $20\frac{1}{5}$ _____ $20\frac{20}{100}$

11. $3\frac{5}{7}$ _____ $4\frac{2}{3}$

12. $19\frac{1}{2}$ _____ $19\frac{9}{18}$

13. $13\frac{3}{5}$ _____ $13\frac{7}{15}$

14. $7\frac{28}{32}$ _____ $7\frac{56}{64}$

15. $3\frac{1}{4}$ _____ $3\frac{3}{8}$

16. $8\frac{2}{9}$ _____ $3\frac{15}{18}$

17. $2\frac{3}{8}$ _____ $2\frac{8}{24}$

18. $5\frac{8}{24}$ _____ $5\frac{3}{24}$

Solve the following problems.

19. Nita weighs $78\frac{3}{4}$ pounds. Alma weighs $78\frac{7}{16}$ pounds. Which girl weighs more?

_____

20. The distance shown on a map is $6\frac{6}{10}$ miles. The actual distance is $6\frac{4}{5}$ miles. Are the distances the same?

_____

Circle the correct answer.

21. Which fraction has the same value as $1\frac{1}{2}$?

   (1) $\frac{10}{8}$ (2) $\frac{11}{8}$

   (3) $\frac{12}{8}$ (4) $\frac{14}{8}$

   (5) $\frac{2}{2}$

22. Which whole number is needed to make $1 = \frac{?}{3}$ true?

   (1) 2   (2) 3

   (3) 4   (4) 5

   (5) none of these

23. $5\frac{3}{4}$ changed to an improper fraction is

(1) $\frac{5}{4}$ (2) $\frac{8}{4}$

(3) $\frac{12}{4}$ (4) $\frac{23}{4}$

(5) none of these

24. $\frac{19}{12}$ changed to a mixed number is

(1) $\frac{7}{12}$ (2) $\frac{12}{19}$

(3) $1\frac{1}{2}$ (4) $1\frac{7}{12}$

(5) none of these

25. $\frac{36}{10}$ expressed in the simplest form is

(1) $\frac{18}{10}$ (2) $\frac{18}{5}$

(3) $\frac{36}{10}$ (4) $3\frac{3}{5}$

(5) none of these

26. $\frac{16}{3}$ expressed in the lowest form is

(1) $4\frac{1}{3}$ (2) $4\frac{3}{3}$

(3) $5\frac{1}{3}$ (4) $5\frac{2}{3}$

(5) $4\frac{4}{3}$

27. Which whole number is needed to make $6 = \frac{48}{?}$ true?

(1) 2   (2) 4

(3) 6   (4) 8

(5) 1

28. Which of the following has the same value as $3\frac{3}{4}$?

(1) $\frac{4}{10}$ (2) $\frac{4}{15}$

(3) $\frac{10}{4}$ (4) $\frac{15}{4}$

(5) none of these

29. Which fraction has the same value as $\frac{18}{36}$?

(1) $\frac{1}{3}$ (2) $\frac{2}{4}$

(3) $\frac{3}{8}$ (4) $\frac{5}{8}$

(5) none of these

30. $\frac{28}{5}$ changed to a mixed number is

(1) $\frac{5}{28}$ (2) $3\frac{3}{5}$

(3) $5\frac{3}{5}$ (4) 6

(5) $5\frac{4}{8}$

31. $\frac{30}{9}$ expressed in the simplest form is

(1) $1\frac{3}{10}$ (2) $2\frac{8}{9}$

(3) $3\frac{1}{3}$ (4) $3\frac{3}{9}$

(5) 3

32. $1\frac{7}{9}$ changed to an improper fraction is

(1) $\frac{10}{9}$ (2) $\frac{11}{9}$

(3) $\frac{15}{9}$ (4) $\frac{16}{9}$

(5) $\frac{17}{9}$

Improper Fractions

A **proper** fraction is a fraction whose numerator is *smaller* than its denominator.

Proper fractions: $\frac{5}{6}$ $\frac{2}{9}$

5 is smaller than 6.
2 is smaller than 9.

A fraction is **improper** if its numerator is *equal to or larger than its denominator.*

Improper fractions: $\frac{6}{5}$ $\frac{4}{3}$ $\frac{7}{7}$

6 is larger than 5.
4 is larger than 3.
7 is equal to 7.

To change an improper fraction to a mixed number, follow these steps:

Step 1 Divide the numerator by the denominator.

Step 2 Reduce the answer to lowest terms.

Example

Change $\frac{18}{8}$ to a mixed number.

Step 1 Divide 18 by 8.
8 goes into 18 two times with a remainder of 2, so

$$\begin{array}{r} 2 \\ 8\overline{)18} \\ -16 \\ \hline 2 \end{array} \qquad \frac{18}{8} = 2\frac{2}{8}$$

Step 2 Reduce the answer to lowest terms whenever necessary.

$$\frac{18}{8} = 2\frac{2}{8} = 2\frac{1}{4}$$

Practice

Change the following to mixed or whole numbers.

1. $\frac{13}{8}$ _____

2. $\frac{24}{15}$ _____

3. $\frac{20}{8}$ _____

4. $\frac{64}{15}$ _____

5. $\frac{21}{6}$ _____

6. $\frac{35}{18}$ _____

7. $\frac{25}{9}$ _____

8. $\frac{17}{3}$ _____

9. $\frac{121}{49}$ _____

10. $\frac{64}{12}$ _____

11. $\frac{11}{3}$ _____

12. $\frac{18}{6}$ _____

13. $\frac{18}{14}$ _____

14. $\frac{50}{49}$ _____

To change a mixed number to an improper fraction, follow these steps:

Step 1 Multiply the whole number by the denominator of the fraction.

Step 2 Add the numerator of the fraction to this product.

Step 3 The sum is the numerator of the improper fraction.

Step 4 The denominator remains the same.

To change a whole number to an improper fraction, follow these steps:

Step 1 Choose any denominator.

Step 2 The numerator will be the product of the denominator and the whole number.

A. Change $2\frac{1}{4}$ to an improper fraction.

Step 1 $2 \times 4 = 8$

Step 2 $8 + 1 = 9$

Step 3 9 is the numerator.

Step 4 4 is the denominator.

$$2\frac{1}{4} = \frac{9}{4}$$

B. Change a whole number, 8, to an improper fraction.

$$8 = \frac{8 \times 10}{10} = \frac{80}{10}$$

$$8 = \frac{8 \times 6}{6} = \frac{48}{6}$$

Practice

Change the following to improper fractions.

15. $3\frac{2}{5}$ _____

16. $4\frac{1}{8}$ _____

17. $9\frac{7}{8}$ _____

18. $8\frac{2}{7}$ _____

19. $1 = \frac{?}{15}$ _____

20. $2\frac{2}{9}$ _____

21. $7\frac{5}{8}$ _____

22. $3\frac{1}{10}$ _____

23. $4\frac{5}{6}$ _____

24. $14 = \frac{?}{9}$ _____

25. $15\frac{2}{3}$ _____

26. $3\frac{1}{8}$ _____

Writing Quotients as Mixed Numbers

The **quotient** is the answer to a division problem. When the answer includes a remainder, one way to write the quotient is as a mixed number.

Example

Divide 179 by 6.

$$
6\overline{)179} \quad 29\tfrac{5}{6}
$$

$$
\begin{array}{r}
29\tfrac{5}{6} \\
6\overline{)179} \\
-12 \\
\hline
59 \\
-54 \\
\hline
5
\end{array}
$$

Divide the whole number 6 into 179.

Take the remainder of 5 and place it over the divisor of 6, making the fraction $\frac{5}{6}$. Write the fraction in the quotient.

The quotient is the mixed number $29\frac{5}{6}$.

Practice

Divide. Write the quotients as mixed numbers. Show your work.

1. $7\overline{)78}$

2. $3\overline{)52}$

3. $4\overline{)49}$

4. $10\overline{)236}$

5. $12\overline{)388}$

6. $14\overline{)45}$

7. $6\overline{)208}$

8. $9\overline{)65}$

9. $8\overline{)156}$

Problem Solving—Mixed Numbers

The steps you have learned to solve word problems can be used with word problems that deal with mixed numbers. Use the following steps:

Step 1   Read the problem and underline the key words. These words usually relate to some mathematics reasoning computation.

Step 2   Ask yourself, Should I add, subtract, multiply, divide, round, or compare? You may have to do more than one of these operations for the same problem.

Step 3   Find the solution. Use your math knowledge to find your answer.

Step 4   Check your answer. Ask yourself, Is the answer reasonable? Did you find what you were asked for?

Some compare words that you should know are as follows:

smaller	larger	equal	less
smaller than	larger than	equal to	more

Example

A. Jackson Real Estate was offering two building lots for sale. Lot A contained $2\frac{3}{5}$ acres and Lot B was $2\frac{4}{5}$ acres. Which lot for sale is larger?

Step 1   Determine which lot is larger. The key word is **larger.**

Step 2   The key word **larger** indicates which operation should occur—comparison. To compare mixed numbers, remember the steps to follow that appear on page 118.

Step 3   Find the solution.

    a. Compare the whole numbers. $2 = 2$

    b. Compare the fractions. $\frac{4}{5} > \frac{3}{5}$

    Therefore, Lot B is larger than Lot A.

Step 4   Check the answer. Does it make sense that $2\frac{4}{5}$ is larger than $2\frac{3}{5}$? Yes, the answer is reasonable.

1. A suit pattern calls for $4\frac{5}{8}$ yards of material. Leroy bought $4\frac{3}{4}$ yards. Did he buy enough material?

_____

2. Maggie put $\frac{3}{4}$ cup of flour into a cake batter. The recipe calls for $\frac{1}{2}$ cup of flour. Did she add more or less flour than the recipe called for?

_____

3. Greg is $5\frac{6}{7}$ feet tall. His brother, Sam, is $5\frac{3}{4}$ feet tall. Who is taller?

_____

4. Caitlin owns a bookstore. This week about $\frac{1}{2}$ of the books sold were fiction, and $\frac{1}{4}$ of the books sold were nonfiction. Were more fiction or nonfiction books sold this week?

_____

5. Nan lives $7\frac{1}{2}$ miles from school. Louie lives $7\frac{1}{8}$ miles from school. Who lives closer to the school?

_____

6. Chuck weighs $185\frac{3}{4}$ pounds. For his weight class, he must weigh no more than $185\frac{1}{2}$ pounds. Does he weigh more or less than the maximum weight?

_____

7. Sadie collects miniature trains, trucks, and cars. Her collection is made up of $\frac{1}{8}$ trains and $\frac{1}{3}$ cars. Does she have more cars or trains in her collection?

_____

8. David mixed a special cleaning solution. The solution called for $1\frac{2}{3}$ cups of liquid soap and $1\frac{2}{8}$ cups of water. Does the solution need more liquid soap or more water?

_____

Write a mixed number for each picture.

1.

_____

2.

_____

Compare the following mixed numbers using >, <, or =.

3. $4\frac{5}{10}$ _____ $4\frac{5}{100}$ **4.** $11\frac{2}{3}$ _____ $11\frac{3}{4}$ **5.** $3\frac{4}{6}$ _____ $3\frac{8}{12}$ **6.** $28\frac{2}{5}$ _____ $28\frac{1}{6}$

Change the following improper fractions to mixed or whole numbers.

7. $\frac{116}{5}$ _____   **8.** $\frac{64}{4}$ _____   **9.** $\frac{39}{2}$ _____   **10.** $\frac{63}{7}$ _____

Change the following mixed numbers to improper fractions.

11. $3\frac{5}{9}$ _____   **12.** $5\frac{5}{8}$ _____   **13.** $19\frac{1}{2}$ _____   **14.** $6\frac{3}{10}$ _____

Divide. Write the quotients as mixed numbers.

15. $8\overline{)2380}$ **16.** $4\overline{)862}$ **17.** $72\overline{)7236}$ **18.** $70\overline{)7030}$

Solve the following problems. Circle the correct answer.

19. If it took 15 days to read a book with 737 pages, how many pages were read daily?

(1) $47\frac{2}{3}$ **(2)** $48\frac{4}{5}$

(3) $49\frac{2}{15}$ **(4)** 46

(5) $46\frac{2}{5}$

20. A piece of twine 438 inches long is cut into 4 equal pieces. How long is each piece?

(1) $19\frac{1}{2}$ in.   **(2)** $109\frac{1}{2}$ in.

(3) $119\frac{1}{2}$ in   **(4)** $190\frac{1}{2}$ in.

(5) $199\frac{1}{2}$ in.

Multiplying and Dividing Fractions and Mixed Numbers

Multiply or divide. Reduce the answers to the lowest terms.

1. $\frac{5}{6} \times \frac{1}{4}$ _____

2. $\frac{3}{8} \times \frac{10}{12}$ _____

3. $\frac{3}{4} \times \frac{8}{9}$ _____

4. $3\frac{7}{8} \times 18$ _____

5. $16 \times \frac{3}{4}$ _____

6. $3\frac{1}{7} \times 1\frac{3}{11}$ _____

7. $3\frac{1}{2} \times 3\frac{1}{2}$ _____

8. $2\frac{3}{4} \times 32$ _____

9. $\frac{1}{4} \div \frac{1}{8}$ _____

10. $12 \div \frac{3}{4}$ _____

11. $3\frac{1}{5} \div 10$ _____

12. $4\frac{5}{6} \div 2\frac{9}{10}$ _____

Solve the following problems. Circle the correct answers.

13. The product of $2\frac{1}{4} \times 2\frac{2}{3} \times 1\frac{1}{2}$ is
    - (1) $3\frac{3}{8}$ (2) 4
    - (3) 6 (4) 9
    - (5) $2\frac{3}{4}$

14. The quotient of $8\frac{1}{6} \div 1\frac{1}{6}$ is
    - (1) $\frac{21}{4}$ (2) $5\frac{1}{4}$
    - (3) 7 (4) $9\frac{19}{36}$
    - (5) 8

15. Ellen bought $22\frac{1}{2}$ pounds of polyfill for pillows. If each pillow needs $2\frac{1}{4}$ pounds, how many pillows can she make?
    - (1) 6 (2) 10
    - (3) 20 (4) $20\frac{1}{4}$
    - (5) $44\frac{3}{4}$

16. Wendell mailed four packages. Each weighed $7\frac{1}{2}$ pounds. What was the total weight of the four packages?
    - (1) $3\frac{1}{2}$ (2) $11\frac{1}{2}$
    - (3) 18 (4) 30
    - (5) 7

Multiplying Fractions by Fractions

The box below is divided into fourths.
Three fourths are shaded.

 $\frac{3}{4}$

To find $\frac{1}{2}$ of $\frac{3}{4}$, divide the boxes in half.

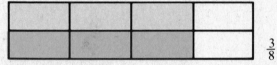 $\frac{3}{8}$

$\frac{1}{2}$ of $\frac{3}{4}$ is shaded darker. The dark shaded part is $\frac{3}{8}$.

$$\frac{1}{2} \text{ of } \frac{3}{4} = \frac{3}{8}$$

Of means **multiply.** So,

$$\frac{1}{2} \times \frac{3}{4} = \frac{3}{8}$$

To multiply a fraction by a fraction (without pictures), follow these steps:

Step 1 Multiply the numerators together.

Step 2 Multiply the denominators together.

Step 3 Reduce to lowest terms.

Examples

A. $\frac{1}{2} \times \frac{3}{4} = \frac{1 \times 3}{2 \times 4} = \frac{3}{8}$

B. $\frac{2}{5} \times \frac{3}{8} = \frac{2 \times 3}{5 \times 8} = \frac{6}{40} = \frac{3}{20}$

Practice

Use each figure to find the product.

1. What is $\frac{1}{2}$ of $\frac{1}{4}$?

2. What is $\frac{1}{3}$ of $\frac{1}{2}$?

3. What is $\frac{1}{4}$ of $\frac{1}{2}$?

Multiply.

4. What is $\frac{2}{3}$ of $\frac{3}{4}$? _____

5. What is $\frac{1}{3}$ of $\frac{1}{3}$? _____

6. What is $\frac{1}{2}$ of $\frac{3}{4}$? _____

Find the product for the following. Show your work.

7. $\frac{1}{4} \times \frac{1}{5}$ _____

8. $\frac{1}{3} \times \frac{2}{3}$ _____

9. $\frac{3}{7} \times \frac{1}{4}$ _____

10. $\frac{7}{12} \times \frac{1}{3}$ _____

11. $\frac{1}{6} \times \frac{1}{4}$ _____

12. $\frac{3}{8} \times \frac{1}{2}$ _____

13. $\frac{1}{2} \times \frac{1}{5}$ _____

14. $\frac{1}{2} \times \frac{1}{6}$ _____

15. $\frac{7}{8} \times \frac{5}{6}$ _____

16. $\frac{1}{8} \times \frac{3}{4}$ _____

17. $\frac{2}{5} \times \frac{4}{7}$ _____

18. $\frac{3}{5} \times \frac{3}{4}$ _____

19. $\frac{1}{3} \times \frac{1}{10}$ _____

20. $\frac{1}{3} \times \frac{1}{4}$ _____

21. $\frac{3}{5} \times \frac{1}{4}$ _____

22. $\frac{1}{5} \times \frac{1}{3}$ _____

23. $\frac{5}{9} \times \frac{2}{3}$ _____

24. $\frac{1}{7} \times \frac{2}{5}$ _____

25. $\frac{1}{4} \times \frac{1}{8}$ _____

26. $\frac{3}{4} \times \frac{5}{8}$ _____

27. $\frac{7}{10} \times \frac{3}{5}$ _____

28. $\frac{4}{5} \times \frac{1}{3}$ _____

29. $\frac{2}{3} \times \frac{4}{5}$ _____

30. $\frac{1}{5} \times \frac{3}{10}$ _____

31. $\frac{6}{7} \times \frac{3}{5}$ _____

32. $\frac{2}{3} \times \frac{7}{9}$ _____

33. $\frac{8}{11} \times \frac{1}{3}$ _____

Cancellation

Cancellation makes multiplication of fractions easier. To cancel, divide any numerator and denominator by the same number. This does not change the value of the answer to the problem.

Study the examples carefully.

Examples

A. Multiply. $\frac{7}{10} \times \frac{10}{11} = ?$

$\frac{7}{\cancel{10}} \times \frac{\cancel{10}^{1}}{11}$

Divide 10 into the numerator. Cross out 10. Write 1 above the 10 as the new numerator.

Divide 10 into the denominator. Cross out 10. Write 1 under the 10 as the new denominator.

$\frac{7}{1} \times \frac{1}{11} = \frac{7 \times 1}{1 \times 11} = \frac{7}{11}$

The problem is now $\frac{7}{1} \times \frac{1}{11}$. Follow the rules for multiplication of fractions.

B. Multiply. $\frac{3}{10} \times \frac{2}{9} = ?$

Cancellation can be used many times in the same problem. Remember to divide one numerator and one denominator by the same number.

$\frac{\cancel{3}^{1}}{\cancel{10}_{5}} \times \frac{\cancel{2}^{1}}{\cancel{9}_{3}}$

Divide 3 and 9 by 3.
Divide 10 and 2 by 2.

$\frac{1}{5} \times \frac{1}{3} = \frac{1 \times 1}{5 \times 3} = \frac{1}{15}$

The problem is now $\frac{1}{5} \times \frac{1}{3}$. Follow the rules for multiplication of fractions.

C. Multiply. $\frac{20}{90} \times \frac{4}{5} = ?$

$\frac{2\cancel{0}}{9\cancel{0}} \times \frac{4}{5}$

Cancel end zeros one for one. This is called **vertical cancellation.**

$\frac{2}{9} \times \frac{4}{5} = \frac{2 \times 4}{9 \times 5} = \frac{8}{45}$

The problem is now $\frac{2}{9} \times \frac{4}{5}$. Follow the rules for multiplication of fractions.

Use cancellation to solve the following problems. Show your work.

1. $\frac{7}{15} \times \frac{3}{14}$

2. $\frac{5}{6} \times \frac{18}{25}$

3. $\frac{9}{16} \times \frac{8}{15}$

4. $\frac{5}{9} \times \frac{3}{5}$

5. $\frac{1}{8} \times \frac{8}{9}$

6. $\frac{11}{21} \times \frac{3}{11}$

7. $\frac{8}{7} \times \frac{14}{24}$

8. $\frac{4}{5} \times \frac{15}{20}$

9. $\frac{5}{6} \times \frac{3}{20}$

10. $\frac{6}{7} \times \frac{7}{8}$

11. $\frac{5}{6} \times \frac{4}{15}$

12. $\frac{1}{6} \times \frac{6}{8}$

13. $\frac{3}{8} \times \frac{16}{21}$

14. $\frac{4}{5} \times \frac{7}{8}$

15. $\frac{3}{8} \times \frac{24}{36}$

16. $\frac{3}{7} \times \frac{7}{9}$

17. $\frac{7}{8} \times \frac{2}{7} \times \frac{1}{2}$

18. $\frac{2}{3} \times \frac{9}{16} \times \frac{5}{6}$

19. $\frac{5}{6} \times \frac{2}{3} \times \frac{9}{15}$

20. $\frac{1}{8} \times \frac{16}{20} \times \frac{4}{5}$

21. $\frac{3}{5} \times \frac{25}{27} \times \frac{3}{5}$

22. $\frac{3}{10} \times \frac{5}{6} \times \frac{2}{3}$

23. $\frac{3}{8} \times \frac{7}{9} \times \frac{4}{5}$

24. $\frac{2}{3} \times \frac{6}{7} \times \frac{14}{15}$

25. $\frac{3}{4} \times \frac{2}{9} \times \frac{1}{2}$

26. $\frac{3}{8} \times \frac{4}{9} \times \frac{3}{4}$

27. $\frac{1}{2} \times \frac{4}{5} \times \frac{5}{8}$

Finding the Majority

In voting, a **simple majority** is $\frac{1}{2}$ the total number of votes plus one. A $\frac{2}{3}$ **majority** is $\frac{2}{3}$ of the total number. A $\frac{3}{4}$ **majority** is $\frac{3}{4}$ of the total number. If the majority is a mixed number, round it to the next whole number. (Don't add one.)

To find the majority of a number, multiply by the fraction needed for the majority.

Find the $\frac{2}{3}$ majority of 66.

$$\frac{2}{\underset{1}{\cancel{3}}} \times \frac{\overset{22}{\cancel{66}}}{1} = \frac{44}{1} = 44$$

Find the majority.

1. Harold Swenson needs a $\frac{2}{3}$ majority to be elected sheriff. If 1,200 people vote, how many votes does he need to be elected? _____

2. Some states change their laws if a $\frac{4}{5}$ majority of persons voting approve. If a state has 5 million persons voting on election day, how many millions will be required to change the law? _____

3. If the United States requires the approval of a $\frac{2}{3}$ majority of the states to amend the Constitution, how many states are needed to make the $\frac{2}{3}$ majority? (There are 50 states.) _____

4. There are 80 members at a block club meeting. To change the bylaws of the club, a $\frac{3}{4}$ majority of the members present must approve. How many persons will make the $\frac{3}{4}$ majority? _____

5. What is the simple majority if there are 584 people voting in an election? _____

Multiplying Fractions by Whole Numbers

Every whole number can be written as a fraction by placing the whole number over the denominator of 1.

8 can be written $\frac{8}{1}$.

The value of 8 does not change because $\frac{8}{1}$ means $8 \div 1$ which is 8.

To multiply a fraction by a whole number, follow these steps.

Step 1   Write the whole number as a fraction.

Step 2   Follow the rules for multiplication of fractions.

Step 3   Reduce to lowest terms.

Examples

A. $8 \times \frac{1}{2} = \frac{\overset{4}{\cancel{8}}}{1} \times \frac{1}{\underset{1}{\cancel{2}}} = \frac{4 \times 1}{1 \times 1} = \frac{4}{1} = 4$

B. $\frac{4}{5} \times 20 = \frac{4}{\underset{1}{\cancel{5}}} \times \frac{\overset{4}{\cancel{20}}}{1} = \frac{4 \times 4}{1 \times 1} = \frac{16}{1} = 16$

C. $36 \times \frac{3}{10} = \frac{\overset{18}{\cancel{36}}}{1} \times \frac{3}{\underset{5}{\cancel{10}}} = \frac{18 \times 3}{1 \times 5} = \frac{54}{5}$

MATH HINT

An example of an improper fraction is $\frac{54}{5}$.

Write it as a mixed number. 5 divided into 54 goes 10 times with a remainder of 4.

$\frac{54}{5} = 10\frac{4}{5}$

Practice

Solve the following. Reduce to lowest terms.

1. $60 \times \frac{7}{15}$ _____

2. $\frac{2}{3} \times 9$ _____

3. $\frac{3}{25} \times 100$ _____

4. $20 \times \frac{1}{4}$ _____

5. $2 \times \frac{1}{3}$ _____

6. $\frac{1}{6} \times 5$ _____

7. $\frac{1}{4} \times 21$ _____

8. $8 \times \frac{1}{6}$ _____

9. $\frac{1}{5} \times 10$ _____

10. $\frac{5}{16} \times 32$ _____

11. $25 \times \frac{7}{10}$ _____

12. $14 \times \frac{5}{7}$ _____

13. $\frac{5}{12} \times 16$ _____

14. $150 \times \frac{1}{3}$ _____

Multiply. Reduce to lowest terms.

15. $3 \times \frac{3}{5}$ _____

16. $\frac{6}{7} \times 5$ _____

17. $\frac{3}{10} \times 19$ _____

18. $48 \times \frac{1}{12}$ _____

19. $\frac{3}{21} \times 6$ _____

20. $\frac{3}{16} \times 32$ _____

21. $36 \times \frac{7}{9}$ _____

22. $\frac{1}{10} \times 150$ _____

23. $\frac{4}{5} \times 15$ _____

24. $\frac{3}{8} \times 36$ _____

25. $\frac{1}{3} \times 39$ _____

26. $90 \times \frac{1}{4}$ _____

27. $\frac{2}{3} \times 18$ _____

28. $\frac{3}{8} \times 20$ _____

29. $54 \times \frac{4}{9}$ _____

30. $50 \times \frac{7}{10}$ _____

31. $8 \times \frac{1}{2}$ _____

32. $\frac{7}{9} \times 63$ _____

33. $\frac{3}{4} \times 18$ _____

34. $\frac{1}{6} \times 42$ _____

35. $64 \times \frac{7}{16}$ _____

36. $\frac{5}{6} \times 12$ _____

37. $800 \times \frac{5}{16}$ _____

38. $235 \times \frac{1}{5}$ _____

39. $\frac{2}{3} \times 32$ _____

40. $30 \times \frac{4}{5}$ _____

41. $27 \times \frac{1}{9}$ _____

42. $\frac{2}{5} \times 15$ _____

Using Measures in Cooking

Dolores works in the bakery as a counter salesperson.
Sometimes she helps the baker prepare the bakery goods. The
tables of measures will help you solve the multiplication
problems below.

Liquid measures
16 ounces (oz.) = 1 pint (pt.)
2 cups = 1 pint
2 pints = 1 quart (qt.)
4 quarts = 1 gallon (gal.)

Dry measures
8 ounces (oz.) = 1 cup
16 ounces = 1 pound (lb.)
12 items = 1 dozen (doz.)

A customer wanted to buy $\frac{1}{4}$ pound of cookies. How many
ounces did she want?

1 pound = 16 ounces

$$\frac{1}{4} \text{ of } 16 = \frac{1}{\underset{1}{4}} \times \frac{\overset{4}{16}}{1} = 4 \text{ ounces}$$

**Solve the following problems. Reduce the answers to lowest
terms.**

1. While preparing a frosting mixture, Dolores used $\frac{1}{2}$ cup of milk. How many ounces did she use?

   _____

2. A bread recipe requires $\frac{2}{3}$ pint of milk. How many ounces will be used?

   _____

3. Dolores sold $\frac{3}{4}$ dozen doughnuts to one customer. How many doughnuts did the customer buy?

   _____

Multiplying Mixed Numbers

To multiply a mixed number by a fraction, a whole number, or another mixed number, follow these steps:

Step 1 Write the mixed numbers as improper fractions.

Step 2 Follow the rules for the multiplication of fractions.

Examples

A. $6 \times 1\frac{1}{2} = \frac{6}{1} \times \frac{3}{2}$

$\overset{3}{\frac{6}{1}} \times \frac{3}{\underset{1}{2}} = \frac{3 \times 3}{1 \times 1} = \frac{9}{1} = 9$

B. $3\frac{1}{2} \times \frac{2}{5} = \frac{7}{2} \times \frac{2}{5}$

$\frac{7}{\underset{1}{2}} \times \overset{1}{\frac{2}{5}} = \frac{7 \times 1}{1 \times 5} = \frac{7}{5} = 1\frac{2}{5}$

C. $5\frac{1}{2} \times 3\frac{1}{3} = \frac{11}{2} \times \frac{10}{3}$

$\frac{11}{\underset{1}{2}} \times \overset{5}{\frac{10}{3}} = \frac{11 \times 5}{1 \times 3} = \frac{55}{3} = 18\frac{1}{3}$

Practice

Multiply.

1. $5\frac{1}{3} \times 3\frac{3}{4}$ _____

2. $6 \times 1\frac{1}{18}$ _____

3. $7 \times 4\frac{5}{14}$ _____

4. $5\frac{2}{5} \times \frac{1}{4}$ _____

5. $8\frac{1}{3} \times 2\frac{7}{10}$ _____

6. $6\frac{1}{4} \times 2\frac{2}{5}$ _____

7. $6\frac{3}{4} \times \frac{1}{3}$ _____

8. $8 \times 5\frac{1}{2}$ _____

9. $15\frac{1}{4} \times 12$ _____

10. $1\frac{15}{16} \times \frac{1}{4}$ _____

11. $6 \times 3\frac{2}{3}$ _____

Multiply.

12. $9\frac{3}{8} \times 72$ _____ **13.** $3\frac{1}{6} \times \frac{2}{3}$ _____

14. $2\frac{3}{5} \times 1\frac{1}{6}$ _____ **15.** $1\frac{7}{8} \times 3\frac{1}{3}$ _____

16. $\frac{1}{3} \times 3\frac{1}{3}$ _____ **17.** $1\frac{1}{5} \times 4\frac{3}{8}$ _____

18. $\frac{2}{3} \times 4\frac{1}{2}$ _____ **19.** $25 \times 8\frac{3}{5}$ _____

20. $4\frac{1}{2} \times 2\frac{1}{3}$ _____ **21.** $2\frac{1}{3} \times 6\frac{1}{2}$ _____

22. $1\frac{2}{3} \times 81$ _____ **23.** $19 \times 1\frac{1}{2}$ _____

24. $2\frac{2}{3} \times \frac{3}{8}$ _____ **25.** $2\frac{1}{4} \times 1\frac{2}{5}$ _____

26. $5\frac{5}{8} \times 3\frac{1}{5}$ _____ **27.** $8 \times 1\frac{5}{6}$ _____

28. $20 \times 5\frac{1}{2}$ _____ **29.** $\frac{1}{6} \times 3\frac{2}{5}$ _____

30. $3\frac{1}{2} \times 2\frac{1}{3}$ _____ **31.** $\frac{2}{3} \times 8\frac{2}{5}$ _____

32. $1\frac{3}{8} \times 24$ _____ **33.** $4\frac{1}{2} \times 6$ _____

34. $\frac{1}{3} \times 5\frac{1}{2}$ _____ **35.** $7\frac{1}{2} \times 2\frac{2}{5}$ _____

36. $2\frac{1}{4} \times 2\frac{2}{3} \times 1\frac{1}{2}$ _____ **37.** $\frac{3}{10} \times 3\frac{1}{8} \times 3\frac{1}{5}$ _____

38. $3\frac{3}{4} \times 1\frac{1}{7} \times 3\frac{1}{3}$ _____ **39.** $1\frac{1}{2} \times \frac{4}{5} \times 2\frac{1}{6}$ _____

Solve the following problems.

40. Harold has a $\frac{1}{4}$-acre garden. He plants one half of his garden with tomatoes. What part of the garden is tomatoes?

        _____

41. If you slept $\frac{1}{4}$ of the day, how many hours would you have slept? (Hint: 24 hours in a day)

        _____

42. There were 303 cars on Level A of the parking garage. If $\frac{2}{3}$ of the cars were U.S. made, how many were foreign imports?

        _____

43. At the speed of $2\frac{3}{4}$ miles per hour, how far will Louise jog in four and a half hours?

        _____

44. Each time a delivery truck makes its work run, it travels $35\frac{3}{4}$ miles. What was the total distance for twenty working days?

        _____

45. A-1 Catering has to prepare 100 sandwiches. Each sandwich will have $3\frac{1}{2}$ ounces of meat and 1 ounce of cheese. How many ounces of meat and cheese will be needed?

        _____

46. Find the cost of $4\frac{1}{2}$ yards of material at $5 a yard. Add to this cost 3 spools of thread at 69 cents each.

        _____

47. A TV repairperson charges $28 per hour for labor costs. If it takes $3\frac{3}{4}$ hours to do a job, what will be the labor cost?

        _____

48. Four families each collected $6\frac{3}{4}$ bags of aluminum cans to be recycled. How many bags of cans did the four families collect?

        _____

49. A muffin recipe calls for $2\frac{1}{4}$ cups of all purpose flour. If the recipe is to be increased by $2\frac{1}{2}$ times, how much flour will be needed?

        _____

Increasing Recipes

It is easy to increase recipes by multiplying by 2, 3, 4, and so on. If a recipe makes 4 servings, multiplying by 2 makes 8 servings. Of course, all the measures of the ingredients must be multiplied by 2. Increase the following recipes. First note the number of servings needed.

Cherry cobbler/1 cobbler
Increase to 6 cobblers

_36_	6 cups pitted tart red cherries
_____	$1\frac{3}{4}$ cups sugar
_____	$\frac{1}{2}$ teaspoon grated lemon rind
_____	$\frac{1}{8}$ teaspoon salt
_____	$\frac{1}{2}$ pkg. piecrust mix
_____	2 tablespoons butter

Yeast rolls/2 dozen rolls
Increase to 48 rolls

_1_	$\frac{1}{2}$ cup milk
_____	2 pkg. dry yeast
_____	$\frac{1}{2}$ cup warm water
_____	$\frac{3}{4}$ cup butter or margarine
_____	$\frac{1}{2}$ cup sugar
_____	4 egg yolks
_____	$4\frac{1}{2}$ cups flour

Banana Bread/2 loaves
Increase to 6 loaves

_____	$1\frac{1}{4}$ pounds of bananas
_____	2 eggs
_____	$\frac{1}{4}$ cup vegetable oil
_____	$\frac{3}{4}$ cup sugar
_____	$\frac{3}{4}$ cup shredded bran cereal
_____	2 teaspoons baking powder

_____	$\frac{1}{2}$ teaspoon salt
_____	$\frac{1}{2}$ teaspoon baking soda
_____	$1\frac{1}{4}$ cups flour
_____	$\frac{1}{2}$ cup shredded coconut
_____	$\frac{1}{3}$ cup walnuts
_____	$\frac{1}{8}$ cup pecans

Dividing Fractions by a Fraction

Fractional values can be divided just as you divide whole numbers and decimals. To divide a fraction by a fraction, **invert the divisor and multiply.** The **divisor** is the **second number.**

You invert this way:

Change $\frac{2}{5}$ to $\frac{5}{2}$, $\frac{6}{11}$ to $\frac{11}{6}$, $\frac{4}{7}$ to $\frac{7}{4}$.

To invert means the numerator becomes the denominator, and the denominator the numerator. The two fractions are called **reciprocals.**

Example

$\frac{3}{5} \div \frac{4}{5} = ?$

To divide a fraction by a fraction, follow these steps:

Step 1   Invert the divisor. Change the division sign to a multiplication sign.

$\frac{3}{5} \times \frac{5}{4}$

Step 2   Cancel whenever possible.

$\frac{3}{5} \times \frac{\overset{1}{5}}{4}$
_{1}

Step 3   Follow the rules for multiplication of fractions.

$\frac{3}{1} \times \frac{1}{4} = \frac{3 \times 1}{1 \times 4} = \frac{3}{4}$

Practice

Divide. Reduce the answers to lowest terms.

1. $\frac{3}{5} \div \frac{1}{2}$ _____

2. $\frac{5}{8} \div \frac{3}{4}$ _____

3. $\frac{7}{8} \div \frac{7}{8}$ _____

4. $\frac{7}{6} \div \frac{2}{3}$ _____

5. $\frac{3}{4} \div \frac{1}{4}$ _____

6. $\frac{2}{3} \div \frac{1}{3}$ _____

7. $\frac{3}{8} \div \frac{5}{16}$ _____

8. $\frac{2}{3} \div \frac{4}{5}$ _____

9. $\frac{3}{10} \div \frac{3}{8}$ _____

10. $\frac{5}{12} \div \frac{5}{8}$ _____

11. $\frac{1}{5} \div \frac{3}{10}$ _____

12. $\frac{3}{4} \div \frac{3}{5}$ _____

13. $\frac{4}{5} \div \frac{1}{5}$ _____

14. $\frac{1}{3} \div \frac{1}{9}$ _____

15. $\frac{3}{5} \div \frac{1}{4}$ _____

16. $\frac{2}{3} \div \frac{1}{8}$ _____

Dividing Fractions by Whole Numbers

Whole numbers can also be inverted. Make the whole number an improper fraction by placing the whole number over 1. Then invert.

Example

$$\frac{8}{11} \div 6 = ?$$

To divide a fraction by a whole number, follow these steps:

Step 1   Change the whole number to an improper fraction. $\frac{8}{11} \div \frac{6}{1}$

Step 2   Invert the divisor. Change the division sign to a multiplication sign. Cancel whenever possible. $\frac{\overset{4}{8}}{11} \times \frac{1}{\underset{3}{6}}$

Step 3   Follow the rules for multiplication of fractions. $\frac{4}{11} \times \frac{1}{3} = \frac{4 \times 1}{11 \times 3} = \frac{4}{33}$

Practice

Divide. Reduce the answers to lowest terms.

1. $\frac{4}{5} \div 8$ _____

2. $\frac{6}{10} \div 3$ _____

3. $\frac{4}{9} \div 8$ _____

4. $\frac{9}{10} \div 6$ _____

5. $\frac{1}{4} \div 7$ _____

6. $\frac{2}{3} \div 8$ _____

7. $\frac{5}{9} \div 4$ _____

8. $\frac{2}{3} \div 7$ _____

9. $\frac{15}{16} \div 9$ _____

10. $\frac{2}{5} \div 6$ _____

11. $\frac{5}{6} \div 5$ _____

12. $\frac{1}{2} \div 5$ _____

13. $\frac{4}{5} \div 12$ _____

14. $\frac{3}{4} \div 3$ _____

15. $\frac{7}{9} \div 14$ _____

16. $\frac{5}{6} \div 10$ _____

Dividing Mixed Numbers by Fractions, Whole Numbers, and Other Mixed Numbers

When there are mixed numbers in a division problem, change them to improper fractions. Then invert the divisor. Follow the rules for multiplication of fractions.

Examples

A. $4\frac{2}{3} \div \frac{7}{9} = ?$

To divide by a fraction, follow these steps:

Step 1   Change the mixed number to an improper fraction. $\frac{14}{3} \div \frac{7}{9}$

Step 2   Invert the divisor. Change the division sign to a multiplication sign. Cancel whenever possible. $\frac{\overset{2}{14}}{\underset{1}{3}} \times \frac{\overset{3}{9}}{\underset{1}{7}}$

Step 3   Follow the rules for multiplication of fractions. $\frac{2}{1} \times \frac{3}{1} = \frac{6}{1} = 6$

B. $3\frac{1}{4} \div 13 = ?$

To divide by a whole number, follow these steps:

Step 1   Change the mixed number to an improper fraction. Change the whole number to an improper fraction by placing 1 in the denominator's position. $\frac{13}{4} \div \frac{13}{1}$

Step 2   Invert the divisor. Change the division sign to a multiplication sign. Cancel whenever possible. $\frac{\overset{1}{13}}{4} \times \frac{1}{\underset{1}{13}}$

Step 3   Follow the rules for multiplication of fractions. $\frac{1}{4} \times \frac{1}{1} = \frac{1 \times 1}{4 \times 1} = \frac{1}{4}$

C. $3\frac{1}{3} \div 1\frac{1}{9} = ?$

To divide by a mixed number, follow these steps:

Step 1   Change both mixed numbers to improper fractions. $\frac{10}{3} \div \frac{10}{9}$

Step 2	Invert the divisor. Change the division sign to a multiplication sign. Cancel whenever possible.	$\dfrac{\overset{1}{\cancel{10}}}{\underset{1}{3}} \times \dfrac{\overset{3}{\cancel{9}}}{\underset{1}{10}}$
Step 3	Follow the rules for multiplication of fractions.	$\dfrac{1}{1} \times \dfrac{3}{1} = \dfrac{1 \times 3}{1 \times 1} = \dfrac{3}{1} = 3$

Practice

Divide. Reduce the answers to lowest terms.

1. $1\frac{1}{9} \div 10$ _____

2. $1\frac{4}{5} \div 18$ _____

3. $1\frac{3}{4} \div \frac{1}{4}$ _____

4. $5\frac{5}{6} \div \frac{5}{6}$ _____

5. $2\frac{2}{3} \div \frac{5}{6}$ _____

6. $3\frac{3}{4} \div \frac{9}{10}$ _____

7. $1\frac{7}{8} \div \frac{3}{4}$ _____

8. $3\frac{3}{4} \div \frac{9}{16}$ _____

9. $6\frac{1}{4} \div 2\frac{1}{2}$ _____

10. $4\frac{4}{9} \div 6\frac{2}{3}$ _____

11. $5\frac{1}{10} \div 3\frac{3}{10}$ _____

12. $3\frac{1}{2} \div 2\frac{3}{4}$ _____

13. $8\frac{1}{3} \div 10$ _____

14. $5\frac{2}{5} \div 3$ _____

15. $3\frac{3}{4} \div \frac{5}{8}$ _____

16. $10\frac{1}{2} \div 2\frac{1}{2}$ _____

17. $8\frac{1}{6} \div 1\frac{1}{6}$ _____

18. $8\frac{1}{3} \div 2\frac{1}{2}$ _____

19. $2\frac{2}{3} \div 1$ _____

20. $5\frac{1}{4} \div 3\frac{1}{2}$ _____

21. $1\frac{1}{2} \div \frac{7}{12}$ _____

22. $3\frac{1}{3} \div 10$ _____

23. $4\frac{2}{3} \div 1\frac{1}{7}$ _____

24. $12 \div 2\frac{2}{5}$ _____

25. $5\frac{3}{5} \div 1\frac{3}{4}$ _____

26. $1\frac{5}{9} \div 3\frac{1}{3}$ _____

27. $5\frac{2}{5} \div 1\frac{4}{5}$ _____

28. $8\frac{3}{4} \div 6\frac{2}{3}$ _____

Using >, <, and =, compare the following expressions.

29. $6 \div \frac{3}{4}$ _____ $\frac{5}{6} \div 10$

30. $\frac{1}{3} \times 2\frac{1}{2}$ _____ $\frac{3}{5} \times 1\frac{1}{4}$

31. $8 \times \frac{5}{6}$ _____ $\frac{1}{2} \times 20$

32. $4\frac{1}{8} \div 1\frac{1}{2}$ _____ $1\frac{7}{9} \div 5\frac{1}{3}$

33. $\frac{16}{25} \div \frac{4}{5}$ _____ $\frac{1}{2} \div \frac{1}{3}$

34. $1\frac{1}{4} \times 1\frac{1}{4}$ _____ $1\frac{1}{6} \times 1\frac{5}{7}$

35. $\frac{2}{3} \times 4\frac{1}{5}$ _____ $2\frac{2}{5} \times 1\frac{1}{6}$

36. $\frac{5}{7} \div \frac{1}{35}$ _____ $20 \div \frac{3}{4}$

37. $\frac{1}{10} \div \frac{1}{5}$ _____ $\frac{7}{16} \div \frac{3}{8}$

38. $3\frac{1}{5} \times 25$ _____ $28 \times 3\frac{3}{4}$

39. $\frac{3}{8} \times \frac{1}{2}$ _____ $\frac{11}{12} \times \frac{3}{4}$

40. $6\frac{1}{4} \div 5$ _____ $4 \div \frac{2}{3}$

41. $\frac{7}{8} \div 2$ _____ $3\frac{1}{3} \div \frac{2}{9}$

42. $5\frac{5}{8} \times 3\frac{1}{3}$ _____ $4\frac{1}{2} \times 6$

43. $5\frac{1}{2} \times 20$ _____ $220 \times \frac{1}{2}$

44. $\frac{7}{16} \div \frac{3}{8}$ _____ $\frac{1}{5} \div \frac{1}{5}$

Increasing or Decreasing Recipes

Increase or decrease the following recipes. Use the rules for multiplication and division of fractions.

Make $1\frac{1}{2}$ times the butter sauce recipe.

1. _____ cups brown sugar

2. _____ cups corn syrup

3. _____ tablespoons butter

4. _____ cups cream

5. _____ cups milk

Make $\frac{1}{2}$ of the muffin recipe.

6. _____ tablespoon butter

7. _____ cup milk

8. _____ eggs

9. _____ tablespoons water

10. _____ yeast cakes

11. _____ cups flour

12. _____ cups sugar

Butter Sauce

$1\frac{1}{4}$ cups brown sugar
$\frac{2}{3}$ cup corn syrup
4 tablespoons butter
$\frac{3}{8}$ cup cream
$\frac{3}{8}$ cup milk

Muffins

1 tablespoon butter
$\frac{1}{2}$ cup milk
2 eggs
3 tablespoons water
$\frac{1}{2}$ yeast cake
$2\frac{1}{2}$ cups flour
$\frac{3}{8}$ cup sugar

Problem Solving—Multiplying and Dividing Fractions

The steps you have learned to solve word problems can be used with word problems that deal with fractions. Use the following steps:

Step 1   Read the problem and underline the key words. These words usually relate to some mathematics reasoning computation.

Step 2   Ask yourself, Should I add, subtract, multiply, divide, round, or compare? You may have to do more than one of these operations for the same problem.

Step 3   Find the solution. Use your math knowledge to find your answer.

Step 4   Check the answer. Ask yourself, Is the answer reasonable? Did you find what you were asked for?

Examples

A. A microwave oven usually sells for $192. It is on sale for $\frac{3}{4}$ of its regular price. What is the sale price?

Step 1   Determine the sale price. The key words are **on sale for $\frac{3}{4}$ of its regular price.**

Step 2   The key words tell you that the regular price is a whole number and the sale price is $\frac{3}{4}$ of it. Since **of** means to multiply, you will be multiplying a whole number by a fraction. Follow the steps you learned on page 133.

Step 3   Find the solution.
    a. Write the whole number as a fraction.

    $$\frac{192}{1}$$

    b. Follow the rules for multiplication of fractions, and reduce to lowest terms.

    $$\overset{48}{\cancel{\frac{192}{1}}} \times \frac{3}{\underset{1}{\cancel{4}}} = \$144$$

Step 4   Check the answer. Does it make sense that $\frac{3}{4}$ of $192 is $144? Yes, the answer is reasonable.

B. How many $\frac{1}{4}$-pound burgers can be made from 5 pounds of meat?

Step 1 Determine how many burgers can be made. The key words are **how many** and **from 5 pounds.**

Step 2 The key words tell you that you have 5 pounds of meat that must be divided into fourths. You will be dividing a whole number by a fraction. Follow the steps you learned on page 133.

Step 3 Find the solution.
    a. Write the whole number as a fraction.

$$\frac{5}{1} \div \frac{1}{4}$$

    b. Invert the divisor, change the division sign to multiplication, and cancel whenever possible.

$$\frac{5}{1} \times \frac{4}{1} = 20 \text{ hamburgers}$$

Step 4 Check the answer. Does it make sense that 20 hamburgers of $\frac{1}{4}$-pound weight can be made from 5 pounds of meat? Yes, the answer is reasonable.

Problem Solving

Solve the following problems.

1. Cindy and Kathleen plan to divide $1\frac{3}{4}$ yards of ribbon equally. What fraction of a yard will each get?

                    _____

2. At the Pak and Ship mail service, $15\frac{3}{4}$ yards of wrapping paper were used to wrap 21 packages. What was the average yardage used per package?

                    _____

3. The distance from Boston to New York City is 222 miles. If a driving trip took $4\frac{1}{5}$ hours, what was the average speed per hour?

                    _____

4. The ACE Moving and Storage Company took $6\frac{3}{4}$ hours to load a truck with furniture from three houses. What was the average amount of time it took for each house?

_____

5. If Brigid cuts $\frac{5}{6}$ of a pie into 5 equal pieces, what fraction of pie will each piece be?

_____

6. If $15\frac{3}{4}$ pounds of grass seeds were used to seed five areas of equal size, how many pounds were used for each area?

_____

7. Shelly rode her bicycle $19\frac{1}{2}$ miles in $2\frac{1}{4}$ hours. What was her average distance per hour?

_____

8. It takes $56\frac{1}{4}$ minutes to wash all of the windows in the Willow Run Office Building. If it takes $3\frac{3}{4}$ minutes to wash one window, how many windows are in the building?

_____

9. The Pennsylvania Turnpike measures 358 miles in length. If the driving rate is 50 miles per hour, how many hours will it take the Mason family to cover the distance? Write answer as a mixed number.

_____

10. Jason is the grill cook at Lopez's Restaurant. He ordered a box of ribeye steaks which weighed 480 ounces. If the average weight of each steak was $7\frac{1}{2}$ ounces, how many steaks were in the box?

_____

Using a Map to Find Distances

1″ = 500 miles

You can use a ruler to help estimate long distances on a map. The scale of the map tells how many miles one inch represents. On the map above, one inch represents 500 miles. To find the distance from Omaha, Nebraska, to Trenton, New Jersey, measure the distance on the map. It is $2\frac{1}{4}$ inches. Now multiply the inches by the mileage scale.

$$2\frac{1}{4} \times 500 = \frac{9}{4} \times \overset{125}{500} = 1125$$

Find the distance.

1. From Salem, Oregon, to Carson City, Nevada _____

2. From Tucson, Arizona, to Albany, New York _____

3. From Helena, Montana, to Madison, Wisconsin _____

4. From Little Rock, Arkansas, to Mobile, Alabama _____

Multiply or divide. Reduce the answers to the lowest terms.

1. $\frac{5}{12} \times \frac{1}{5}$ _____

2. $\frac{14}{15} \times \frac{3}{7}$ _____

3. $\frac{7}{9} \times 36$ _____

4. $9\frac{3}{4} \times 1\frac{1}{3}$ _____

5. $5\frac{3}{7} \times 2\frac{5}{8}$ _____

6. $9 \times \frac{3}{8}$ _____

7. $1\frac{1}{15} \times \frac{3}{8}$ _____

8. $\frac{1}{6} \div \frac{5}{9}$ _____

9. $\frac{1}{3} \div 9$ _____

10. $2\frac{3}{4} \div 2\frac{1}{7}$ _____

11. $6\frac{4}{5} \div 2\frac{2}{5}$ _____

12. $4 \div \frac{3}{4}$ _____

13. $4\frac{1}{8} \div 5\frac{1}{2}$ _____

14. $1\frac{7}{9} \div 5\frac{1}{3}$ _____

15. $1\frac{3}{5} \div 3\frac{1}{5}$ _____

16. $1\frac{3}{4} \div 1\frac{2}{3}$ _____

17. $2\frac{1}{5} \div 10$ _____

18. $\frac{1}{6} \div \frac{7}{8}$ _____

Solve the following problems.

19. Find the quotient of $3\frac{2}{3} \div \frac{11}{12}$.

_____

20. Find the product of $2\frac{2}{3} \times \frac{4}{5} \times \frac{5}{6}$.

_____

21. Leslie had $2\frac{7}{8}$ pounds of cookies. If she divides the cookies into 4 equal packages, what is the weight of each package?

_____

22. Thirty-five men each worked $7\frac{1}{5}$ hours for the fire department. How many hours did they work altogether?

_____

Adding and Subtracting Fractions and Mixed Numbers

Add or subtract. Reduce answers to lowest terms.

1. $\frac{13}{18}$
   $+\frac{17}{18}$

2. $\frac{13}{20}$
   $-\frac{8}{20}$

3. $\frac{5}{8}$
   $+\frac{1}{5}$

4. $7\frac{1}{6}$
   $-\frac{5}{6}$

5. $7\frac{3}{8}$
   $+5\frac{1}{6}$

6. $18\frac{2}{5}$
   $-16\frac{4}{5}$

7. 32
   $-\ 8\frac{1}{2}$

8. $13\frac{3}{4}$
   $-10\frac{1}{3}$

Solve the following problems.

9. Arturo has boards measuring $3\frac{1}{2}$ feet, $4\frac{1}{4}$ feet, and $20\frac{1}{3}$ feet. If he lays these boards end to end to make a walk, how long will the walk be?

10. Ursula had $3\frac{1}{6}$ yards of silk. She used $1\frac{2}{3}$ yards to make a scarf. How many yards are left?

11. Fred allowed $\frac{1}{3}$ of his income for rent, $\frac{1}{4}$ for food, and $\frac{1}{6}$ for transportation. What fractional part of his income did he allow for these three items?

12. The distance from town A to town B is $39\frac{1}{4}$ miles. If Dave drove $19\frac{5}{6}$ miles towards town B, how far is he from town B?

Adding and Subtracting Like Fractions

Fractions with the same denominator have a **common denominator.**
The fractions $\frac{1}{7}$ and $\frac{6}{7}$ have a common denominator of 7. These
fractions are also called **like fractions.**

To add like fractions, follow these steps:

Step 1   Add only the numerators.

Step 2   Use the same denominator in the
answer.

Step 3   Write the sum in lowest terms.

To subtract like fractions, follow these steps:

Step 1   Subtract only the numerators.

Step 2   Use the same denominator in the
answer.

Step 3   Write the difference in lowest
terms.

Examples

A.
$$\begin{array}{r} \frac{7}{10} \\ +\frac{1}{10} \\ \hline \frac{8}{10} = \frac{4}{5} \end{array}$$

B.
$$\begin{array}{r} \frac{5}{8} \\ +\frac{7}{8} \\ \hline \frac{12}{8} = 1\frac{4}{8} = 1\frac{1}{2} \end{array}$$

C.
$$\begin{array}{r} \frac{7}{10} \\ -\frac{1}{10} \\ \hline \frac{6}{10} = \frac{3}{5} \end{array}$$

D.
$$\begin{array}{r} \frac{1}{5} \\ -\frac{1}{5} \\ \hline 0 \end{array}$$

Practice

Add or subtract. Write the answers in lowest terms.

1.
$$\begin{array}{r} \frac{3}{11} \\ +\frac{6}{11} \\ \hline \end{array}$$

2.
$$\begin{array}{r} \frac{2}{25} \\ +\frac{23}{25} \\ \hline \end{array}$$

3.
$$\begin{array}{r} \frac{3}{8} \\ +\frac{7}{8} \\ \hline \end{array}$$

4.
$$\begin{array}{r} \frac{13}{15} \\ +\frac{1}{15} \\ \hline \end{array}$$

5.
$$\begin{array}{r} \frac{13}{16} \\ -\frac{11}{16} \\ \hline \end{array}$$

6.
$$\begin{array}{r} \frac{8}{10} \\ -\frac{2}{10} \\ \hline \end{array}$$

7.
$$\begin{array}{r} \frac{21}{25} \\ -\frac{16}{25} \\ \hline \end{array}$$

8.
$$\begin{array}{r} \frac{5}{6} \\ -\frac{2}{6} \\ \hline \end{array}$$

9. $\dfrac{13}{20}$
$+ \dfrac{11}{20}$

10. $\dfrac{8}{10}$
$+ \dfrac{4}{10}$

11. $\dfrac{5}{15}$
$+ \dfrac{12}{15}$

12. $\dfrac{3}{8}$
$+ \dfrac{4}{8}$

13. $\dfrac{18}{25}$
$- \dfrac{8}{25}$

14. $\dfrac{21}{50}$
$- \dfrac{11}{50}$

15. $\dfrac{5}{9}$
$- \dfrac{5}{9}$

16. $\dfrac{8}{9}$
$- \dfrac{2}{9}$

17. $\dfrac{19}{20}$
$+ \dfrac{11}{20}$

18. $\dfrac{63}{100}$
$+ \dfrac{27}{100}$

19. $\dfrac{6}{11}$
$+ \dfrac{8}{11}$

20. $\dfrac{13}{16}$
$+ \dfrac{15}{16}$

21. $\dfrac{5}{16}$
$- \dfrac{1}{16}$

22. $\dfrac{11}{12}$
$- \dfrac{5}{12}$

23. $\dfrac{11}{18}$
$- \dfrac{2}{18}$

24. $\dfrac{11}{24}$
$- \dfrac{2}{24}$

25. $\dfrac{3}{4}$
$+ \dfrac{1}{4}$

26. $\dfrac{5}{8}$
$+ \dfrac{7}{8}$

27. $\dfrac{3}{5}$
$+ \dfrac{1}{5}$

28. $\dfrac{5}{12}$
$+ \dfrac{3}{12}$

29. $\dfrac{5}{6}$
$- \dfrac{1}{6}$

30. $\dfrac{10}{12}$
$- \dfrac{2}{12}$

31. $\dfrac{40}{50}$
$- \dfrac{30}{50}$

32. $\dfrac{13}{32}$
$- \dfrac{9}{32}$

Adding and Subtracting Unlike Fractions

Unlike fractions are fractions with different denominators. Fractions such as $\frac{1}{5}$ and $\frac{3}{10}$ are unlike fractions. Before fractions can be added or subtracted, they must have the same denominator.

To add unlike fractions, follow these steps:

Step 1 Find a common denominator.

Step 2 Add the numerators.

Step 3 Write the sum in lowest terms.

To subtract unlike fractions, follow these steps:

Step 1 Find a common denominator.

Step 2 Subtract the numerators.

Step 3 Write the difference in lowest terms.

Examples

A. $\frac{1}{5} + \frac{1}{15}$

$$\begin{array}{r} \frac{1}{5} = \frac{3}{15} \\ + \frac{1}{15} = \frac{1}{15} \\ \hline \frac{4}{15} \end{array}$$

Since $15 = 5 \times 3$, 15 is a common denominator.

B. $\frac{7}{8} - \frac{1}{3}$

$$\begin{array}{r} \frac{7}{8} = \frac{21}{24} \\ - \frac{1}{3} = \frac{8}{24} \\ \hline \frac{13}{24} \end{array}$$

Since 3 does not divide evenly into 8, multiply 3 × 8 to get 24 for a common denominator.

> **MATH HINT**
>
> If the smaller denominator divides evenly into the larger denominator, the larger number is a common denominator. Otherwise, you can multiply the two denominators to get a common denominator. (See Lesson 35.)

Practice

Add or subtract. Write the answers in lowest terms.

1. $\begin{array}{r} \frac{1}{12} \\ + \frac{1}{8} \\ \hline \end{array}$

2. $\begin{array}{r} \frac{2}{9} \\ + \frac{1}{4} \\ \hline \end{array}$

3. $\begin{array}{r} \frac{3}{10} \\ + \frac{5}{6} \\ \hline \end{array}$

4. $\begin{array}{r} \frac{5}{7} \\ + \frac{2}{5} \\ \hline \end{array}$

5. $\dfrac{7}{12}$ $-\dfrac{1}{4}$

6. $\dfrac{3}{5}$ $-\dfrac{4}{10}$

7. $\dfrac{19}{20}$ $-\dfrac{9}{10}$

8. $\dfrac{25}{25}$ $-\dfrac{1}{4}$

9. $\dfrac{1}{5}$ $+\dfrac{3}{8}$

10. $\dfrac{3}{4}$ $+\dfrac{1}{3}$

11. $\dfrac{11}{16}$ $+\dfrac{1}{4}$

12. $\dfrac{11}{18}$ $+\dfrac{5}{6}$

13. $\dfrac{3}{5}$ $-\dfrac{3}{15}$

14. $\dfrac{13}{18}$ $-\dfrac{4}{9}$

15. $\dfrac{14}{15}$ $-\dfrac{8}{10}$

16. $\dfrac{19}{20}$ $-\dfrac{17}{25}$

17. $\dfrac{5}{12}$ $+\dfrac{3}{8}$

18. $\dfrac{5}{8}$ $+\dfrac{5}{12}$

19. $\dfrac{9}{14}$ $+\dfrac{1}{2}$

20. $\dfrac{13}{15}$ $+\dfrac{7}{10}$

21. $\dfrac{6}{7}$ $-\dfrac{2}{3}$

22. $\dfrac{11}{12}$ $-\dfrac{7}{8}$

23. $\dfrac{5}{6}$ $-\dfrac{3}{10}$

24. $\dfrac{4}{5}$ $-\dfrac{1}{2}$

25. $\dfrac{1}{5}$ $\dfrac{1}{4}$ $+\dfrac{1}{2}$

26. $\dfrac{1}{2}$ $\dfrac{2}{9}$ $+\dfrac{1}{6}$

27. $\dfrac{8}{15}$ $\dfrac{1}{2}$ $+\dfrac{5}{6}$

28. $\dfrac{2}{3}$ $\dfrac{3}{4}$ $+\dfrac{1}{8}$

29. $\dfrac{5}{6}$ $-\dfrac{3}{8}$

30. $\dfrac{8}{9}$ $-\dfrac{1}{6}$

31. $\dfrac{9}{12}$ $-\dfrac{3}{4}$

32. $\dfrac{7}{8}$ $-\dfrac{21}{24}$

Adding Like Mixed Numbers

Numbers such as $1\frac{1}{3}$ and $2\frac{2}{3}$ are **like mixed numbers.** Their fractions have common denominators. To add like mixed numbers, follow these steps:

Step 1   Add the whole numbers.
Step 2   Add the fractions.
Step 3   Write the sum in lowest terms.

Examples

A. $1\frac{1}{3}$
$+\,2\frac{2}{3}$
$\overline{3\frac{3}{3}} = 4$

Step 1 $1\;+\;2\;=\;3$
Step 2 $\frac{1}{3}+\frac{2}{3}=\frac{3}{3}$
Step 3 $3\frac{3}{3}=3\;+\;1\;=\;4$

B. $2\frac{1}{4}$
$+\,1\frac{1}{4}$
$\overline{3\frac{2}{4}} = 3\frac{1}{2}$

C. $8\frac{4}{9}$
$+\,6\frac{7}{9}$
$\overline{14\frac{11}{9}} = 15\frac{2}{9}$

D. $10\frac{5}{12}$
$+\;\;\;4\frac{11}{12}$
$\overline{14\frac{16}{12}} = 15\frac{4}{12} = 15\frac{1}{3}$

Practice

Add. Write the sums in lowest terms.

1. $8\frac{3}{4}$
$+\,4\frac{3}{4}$

2. $8\frac{7}{10}$
$+\,1\frac{3}{10}$

3. $7\frac{2}{9}$
$+\,6\frac{5}{9}$

4. $21\frac{11}{18}$
$+\,21\frac{7}{18}$

5. $9\frac{2}{15}$
$+\;\;\;\frac{8}{15}$

6. $3\frac{1}{7}$
$+\,3\frac{3}{7}$

155

7. $3\frac{1}{6}$
   $6\frac{4}{6}$
   $+\ 2\frac{3}{6}$

8. $5\frac{5}{16}$
   $8\frac{7}{16}$
   $+\ 6\frac{9}{16}$

9. $1\frac{3}{8}$
   $\frac{4}{8}$
   $+\ 5\frac{1}{8}$

10. $4\frac{2}{7}$
    $7\frac{3}{7}$
    $+\ 5$

11. $7\frac{19}{20}$
    $+\ 6\frac{1}{20}$

12. $2\frac{7}{12}$
    $+\ 4\frac{3}{12}$

13. $5\frac{5}{10}$
    $+\ 10\frac{3}{10}$

14. $7\frac{3}{11}$
    $+\ 5\frac{8}{11}$

15. $13\frac{1}{3}$
    $+\ 17\frac{2}{3}$

16. $6\frac{7}{9}$
    $+\ 3\frac{5}{9}$

17. $2\frac{8}{10}$
    $+\ 1\frac{4}{10}$

18. $10\frac{5}{12}$
    $+\ 4\frac{11}{12}$

19. $2\frac{5}{8}$
    $4\frac{6}{8}$
    $+\ 3\frac{7}{8}$

20. $3\frac{9}{12}$
    $5\frac{3}{12}$
    $+\ 7\frac{5}{12}$

21. $\frac{6}{10}$
    $3\frac{8}{10}$
    $+\ 1\frac{5}{10}$

22. $6\frac{12}{20}$
    $15\frac{5}{20}$
    $+\ \frac{3}{20}$

23. $4\frac{2}{4}$
    3
    $+\ 1\frac{3}{4}$

24. $7\frac{1}{6}$
    $3\frac{2}{6}$
    $+\ 11\frac{3}{6}$

25. $9\frac{5}{6}$
    $2\frac{4}{6}$
    $+\ 11\frac{1}{6}$

26. $5\frac{1}{5}$
    $2\frac{2}{5}$
    $+\ 13\frac{4}{5}$

Adding Unlike Mixed Numbers

Numbers such as $3\frac{2}{3}$ and $3\frac{1}{2}$ are **unlike mixed numbers.** They have different denominators. To add mixed numbers with different denominators, follow these steps:

Step 1   Find a common denominator.

Step 2   Add the whole numbers.

Step 3   Add the fractions.

Step 4   Write the sum in lowest terms.

Examples

A.
$$3\frac{2}{3} = 3\frac{4}{6}$$
$$+\,3\frac{1}{2} = 3\frac{3}{6}$$
$$\overline{\qquad\qquad 6\frac{7}{6} = 7\frac{1}{6}}$$

Step 1   6 is a common denominator because $3 \times 2 = 6$.

Step 2 $3 + 3 = 6$

Step 3 $\frac{4}{6} + \frac{3}{6} = \frac{7}{6}$

Step 4 $6 + \frac{7}{6} = 6 + 1\frac{1}{6} = 7\frac{1}{6}$

B.
$$5\frac{2}{9} = 5\frac{4}{18}$$
$$+\,2\frac{5}{6} = 2\frac{15}{18}$$
$$\overline{\qquad\quad 7\frac{19}{18} = 8\frac{1}{18}}$$

Practice

Add. Write the sums in lowest terms.

1.
$$4\frac{5}{8}$$
$$11\frac{7}{16}$$
$$+\,13\frac{1}{2}$$

2.
$$2\frac{1}{2}$$
$$3\frac{3}{5}$$
$$+\,4\frac{4}{15}$$

3.
$$13\frac{1}{2}$$
$$10\frac{1}{3}$$
$$+\ \ 5\frac{1}{4}$$

4.
$$7\frac{5}{6}$$
$$3\frac{1}{3}$$
$$+\,8\frac{3}{4}$$

5.
$$3\frac{1}{4}$$
$$+\,1\frac{7}{10}$$

6.
$$83\frac{5}{6}$$
$$+\,29\frac{7}{8}$$

7.
$$65\frac{1}{5}$$
$$+\,33\frac{1}{3}$$

8.
$$5\frac{1}{3}$$
$$+\,6\frac{2}{9}$$

9. $2\frac{5}{9}$
$+6\frac{3}{4}$

10. $10\frac{13}{20}$
$+7\frac{4}{5}$

11. $4\frac{1}{2}$
$+7\frac{2}{5}$

12. $6\frac{5}{8}$
$+4\frac{5}{12}$

13. $7\frac{1}{11}$
$51\frac{1}{2}$
$+4\frac{5}{22}$

14. $90\frac{11}{12}$
$14\frac{1}{6}$
$+7\frac{3}{4}$

15. $3\frac{5}{7}$
$7\frac{2}{3}$
$+6\frac{8}{21}$

16. $9\frac{5}{8}$
$11\frac{3}{16}$
$+4\frac{1}{4}$

Problem Solving

Solve the following problems. Write your answers in lowest terms.

17. Mae crochets scarves. The first day she finished $8\frac{1}{8}$ inches. She crocheted $6\frac{3}{8}$ inches on the second day and $10\frac{1}{4}$ inches on the third day. How long was the scarf?

_____

18. Miguel weighs $153\frac{1}{4}$ pounds. His brother weighs $148\frac{1}{2}$ pounds. What is their combined weight?

_____

19. Marilyn works in a nursery. Two babies weighed $7\frac{1}{4}$ pounds each. A third baby weighed $6\frac{1}{2}$ pounds, and the fourth baby weighed $5\frac{3}{4}$ pounds. What was the total weight of the four babies?

_____

20. Emily has $3\frac{7}{8}$ yards of material in one piece and $6\frac{1}{2}$ yards in another piece. How much material does she have?

_____

21. Benito worked on his car for $1\frac{2}{3}$ hours on Friday and $\frac{3}{4}$ hours on Saturday. How many hours did he work on his car?

_____

22. May bought $\frac{1}{3}$ pound of American cheese, $\frac{1}{4}$ pound of Swiss cheese, and $\frac{1}{2}$ pound of cheddar. What was the total weight of the cheese?

_____

Adding Volunteer Hours

Senior citizens volunteer at Morningside Daycare Center. The chart below records their hours for one week. Find the total hours each volunteer worked. The first answer is completed for you.

Volunteer	Feb. 2	Feb. 3	Feb. 4	Feb. 5	Total hours
1. Simms, T.	$3\frac{1}{4}$	$1\frac{1}{2}$	4	$2\frac{3}{4}$	$11\frac{1}{2}$
2. Aran, B.	$2\frac{2}{3}$	$2\frac{1}{4}$	$1\frac{1}{2}$	$1\frac{3}{4}$	
3. Varca, C.	$2\frac{1}{4}$	$1\frac{2}{3}$	$3\frac{2}{3}$	$2\frac{3}{4}$	
4. Wu, L.	$1\frac{1}{2}$	$3\frac{1}{4}$	2	$4\frac{1}{2}$	
5. Caban, S.	$1\frac{1}{2}$	$2\frac{1}{4}$	$1\frac{3}{4}$	$2\frac{1}{2}$	
6. Rivera, G.	$3\frac{1}{2}$	2	$2\frac{1}{4}$	3	

Simms, T. Find a common denominator for the fractions.

$$3\frac{1}{4} = 3\frac{1}{4}$$
$$1\frac{1}{2} = 1\frac{2}{4}$$
$$4 = 4$$
$$+ 2\frac{3}{4} = 2\frac{3}{4}$$
$$\overline{\qquad 10\frac{6}{4}}$$

Reduce the sum to lowest terms.

$$10\frac{6}{4} = 11\frac{2}{4} = 11\frac{1}{2}$$

Subtracting Like Mixed Numbers

To subtract like mixed numbers, follow these steps:

Step 1 Subtract the fractions, renaming when necessary.

Step 2 Subtract the whole numbers.

Step 3 Write the difference in lowest terms.

Examples

A.
$$7\frac{5}{7}$$
$$-2\frac{4}{7}$$
$$5\frac{1}{7}$$

As $\frac{4}{7}$ is less than $\frac{5}{7}$, it is easy to subtract the fractions.

Step 1 $\frac{5}{7} - \frac{4}{7} = \frac{1}{7}$

Step 2 $7 - 2 = 5$

Step 3 The difference is $5\frac{1}{7}$.

B.
$$10\frac{4}{9} = 9\frac{9}{9} + \frac{4}{9} = 9\frac{13}{9}$$
$$-\ \ 4\frac{5}{9} \qquad\qquad = 4\frac{5}{9}$$
$$5\frac{8}{9}$$

$\frac{5}{9}$ is greater than $\frac{4}{9}$. To subtract, rename the mixed number, $10\frac{4}{9}$. Use the common denominator to write $10\frac{4}{9}$ as $9 + \frac{9}{9} + \frac{4}{9}$. The fractions add to $\frac{13}{9}$. $\frac{13}{9}$ is greater than the subtrahend fraction, $\frac{5}{9}$. Now subtract.

Step 1 $\frac{13}{9} - \frac{5}{9} = \frac{8}{9}$

Step 2 $9 - 4 = 5$

Step 3 The difference is $5\frac{8}{9}$.

C.
$$4 \ \ = 3\frac{4}{4}$$
$$-\ 1\frac{1}{4} = 1\frac{1}{4}$$
$$2\frac{3}{4}$$

4 is a whole number. A fraction cannot be subtracted from a whole number until the whole number is renamed. Use the denominator of the fraction to rename 4 as $3\frac{4}{4}$. Now subtract.

Step 1 $\frac{4}{4} - \frac{1}{4} = \frac{3}{4}$

Step 2 $3 - 1 = 2$

Step 3 The difference is $2\frac{3}{4}$.

Practice

Subtract. Write the answers in lowest terms.

1.
$$2\frac{11}{16}$$
$$-1\frac{5}{16}$$

2.
$$4\frac{1}{18}$$
$$-1\frac{7}{18}$$

3.
$$7\frac{3}{24}$$
$$-2\frac{11}{24}$$

4.
$$8\frac{4}{5}$$
$$-6\frac{1}{5}$$

5. $8\frac{1}{3}$
$-\ 6\frac{2}{3}$

6. $20\frac{13}{18}$
$-\ \ 4\frac{7}{18}$

7. 12
$-\ 4\frac{5}{16}$

8. $8\frac{5}{10}$
$-\ 5\frac{1}{10}$

9. $15\frac{9}{14}$
$-\ \ 2\frac{6}{14}$

10. $23\frac{7}{18}$
$-\ 14\frac{14}{18}$

11. $14\frac{5}{15}$
$-\ \ 8\frac{7}{15}$

12. 34
$-\ \ 6\frac{5}{9}$

13. $13\frac{1}{9}$
$-\ \ 7\frac{4}{9}$

14. $3\frac{11}{12}$
$-\ 1\frac{7}{12}$

15. $12\frac{3}{4}$
$-\ \ 9\frac{1}{4}$

16. $26\frac{9}{20}$
$-\ 11\frac{3}{20}$

17. $13\frac{7}{12}$
$-\ \ 9\frac{5}{12}$

18. 8
$-\ 4\frac{2}{3}$

19. $40\frac{1}{4}$
$-\ 39\frac{3}{4}$

20. $9\frac{5}{8}$
$-\ 4\frac{2}{8}$

21. 23
$-\ 13\frac{15}{25}$

22. $4\frac{2}{8}$
$-\ 1\frac{5}{8}$

23. 15
$-\ \ 7\frac{5}{6}$

24. $6\frac{5}{12}$
$-\ 4$

25. $6\frac{4}{8}$
$-\ 5\frac{5}{8}$

26. $14\frac{3}{14}$
$-\ 11\frac{9}{14}$

27. 15
$-\ \ 8\frac{8}{15}$

28. $4\frac{1}{18}$
$-\ 1\frac{7}{18}$

Subtracting Unlike Mixed Numbers

To subtract mixed numbers with different denominators, follow these steps:

Step 1 Find a common denominator.

Step 2 Subtract the fractions, renaming when necessary.

Step 3 Subtract the whole numbers.

Step 4 Write the difference in lowest terms.

Examples

A.
$$9\frac{4}{5} = 9\frac{20}{25}$$
$$-6\frac{3}{25} = 6\frac{3}{25}$$
$$\overline{\phantom{-6\frac{3}{25} = }\; 3\frac{17}{25}}$$

B. Step 1 / Step 2
$$13\frac{1}{5} = 13\frac{4}{20} = 12\frac{20}{20} + \frac{4}{20} = 12\frac{24}{20}$$
$$-\;\; 2\frac{3}{4} = \;\; 2\frac{15}{20} \qquad\qquad\qquad\qquad 2\frac{15}{20}$$
$$\text{Step 3} \rightarrow 10\frac{9}{20}$$
Step 4

C.
$$15\frac{7}{10} = 15\frac{21}{30}$$
$$-\;\; 8\frac{8}{15} = \;\; 8\frac{16}{30}$$
$$\overline{\phantom{-\;\; 8\frac{8}{15} = }\;\; 7\frac{5}{30} = 7\frac{1}{6}}$$

Practice

Subtract. Write the answers in lowest terms.

1.
$$10\frac{3}{20}$$
$$-\;\; 7\frac{4}{5}$$

2.
$$12\frac{3}{7}$$
$$-\;\; 4\frac{2}{3}$$

3.
$$15\frac{1}{2}$$
$$-\;\; 6\frac{1}{6}$$

4.
$$7\frac{5}{8}$$
$$-\;3\frac{5}{16}$$

5.
$$11\frac{4}{7}$$
$$-\;\; 7\frac{10}{21}$$

6.
$$13\frac{11}{16}$$
$$-\;\; 9\frac{3}{8}$$

7. $7\frac{1}{2}$
$-\ 3\frac{3}{4}$

8. $1\,3\frac{4}{5}$
$-\ \ \ 4\frac{2}{3}$

9. $5\frac{1}{5}$
$-\ 2\frac{3}{4}$

10. $1\,2\frac{5}{16}$
$-\ \ \ 7\frac{1}{2}$

11. $5\,7\frac{1}{3}$
$-\ 2\,3\frac{3}{4}$

12. $4\frac{1}{8}$
$-\ 1\frac{3}{6}$

13. $1\,5\frac{8}{15}$
$-\ \ \ 8\frac{7}{10}$

14. $9\frac{4}{5}$
$-\ 6\frac{3}{25}$

15. $1\,8\frac{3}{10}$
$-\ 1\,3\frac{3}{4}$

16. $7\frac{1}{6}$
$-\ 2\frac{3}{4}$

17. $5\frac{3}{8}$
$-\ 1\frac{7}{12}$

18. $9\frac{5}{16}$
$-\ 2\frac{7}{8}$

19. 6
$-\ 1\frac{3}{4}$

20. $4\frac{2}{3}$
$-\ 3\frac{1}{4}$

21. $6\frac{1}{3}$
$-\ 2\frac{1}{12}$

22. $5\frac{1}{6}$
$-\ 2\frac{1}{3}$

23. $7\frac{7}{12}$
$-\ 3\frac{1}{8}$

24. $8\frac{2}{3}$
$-\ 3\frac{4}{5}$

163

Problem Solving—Adding and Subtracting Fractions

You have already solved addition and subtraction problems with whole numbers and decimals. The steps you followed for those problems can also apply to addition and subtraction problems with fractions. Recall these steps:

Step 1   Read the problem and underline the key words. These words will generally relate to some mathematics reasoning computation.

Step 2   Make a plan to solve the problem. Ask yourself, Should I add, subtract, multiply, divide, round, or compare? You may have to do more than one of these operations for the same problem.

Step 3   Find the solution. Use your math knowledge to find your answer.

Step 4   Check the answer. Ask yourself, Is the answer reasonable? Did you find what you were asked for?

Here is a review of some key words for addition and subtraction.

Addition	Subtraction
altogether	decreased by
both	diminished by
increase	difference
sum	how much less
total	how much more
together	remainder

Examples

A. Michelle had a board $7\frac{3}{4}$ feet long. She cut $1\frac{1}{2}$ feet from the board. How much is left?

Step 1   Determine how much board is left. The key word is **left.**

Step 2   The key word indicates which operation should occur— subtraction. Look carefully at the numbers used in this problem. They are unlike mixed numbers so you must follow the steps you learned on page 162.

Step 3   Find the solution.

    a. Find a common denominator for $\frac{1}{2}$ and $\frac{3}{4}$.

$$1 \times 2 = 2 \qquad 1 \times 4 = 4$$
$$2 \times 2 = 4$$

The common denominator is 4

    b. Subtract the fractions, the whole numbers, and write the difference in lowest terms.

$$7\,\tfrac{3}{4} = \quad 7\,\tfrac{3}{4}$$
$$-1\,\tfrac{1}{2} = -1\,\tfrac{2}{4}$$
$$\overline{\qquad\qquad 6\,\tfrac{1}{4}\ \text{feet left}}$$

Step 4   Check the answer. Does it make sense that $6\frac{1}{4}$ feet would be left? Yes, the answer is reasonable.

B. Catherine had a bolt of fabric $43\frac{1}{8}$ yards long. She purchased $22\frac{5}{8}$ yards more. How many yards does she have altogether?

Step 1   Determine how many yards of material she has altogether. The key word is **altogether.** This word can also mean sum or total.

Step 2   The key word indicates which operation should occur—addition. Look carefully at the numbers used in this problem. They are like mixed numbers so you must follow the steps you learned on page 155.

Step 3   Find the solution.

    a. Add the whole numbers.

$$4\,3\,\tfrac{1}{8}$$
$$+2\,2\,\tfrac{5}{8}$$
$$\overline{6\,5\qquad}$$

    b. Add the fractions and write the sum in lowest terms.

$$4\,3\,\tfrac{1}{8}$$
$$+2\,2\,\tfrac{5}{8}$$
$$\overline{6\,5\,\tfrac{6}{8} = 65\,\tfrac{3}{4}\ \text{yards}}$$

Step 4   Check the answer. Does it make sense that $65\frac{3}{4}$ yards would be the sum or total? Yes, the answer is reasonable.

Solve the following problems.

1. Lloyd has two boards, one $\frac{3}{4}$ inch thick and the other $\frac{3}{8}$ inch thick. The first board is how much thicker? _____

2. A wire $4\frac{5}{6}$ yards long broke off $1\frac{1}{3}$ yards from one end. How many yards of wire were left? _____

3. Lucas had $5\frac{1}{3}$ gallons of paint. He used $2\frac{2}{3}$ gallons. How much did he have left?

_____

4. A bolt of fabric has $43\frac{7}{8}$ yards. Luanne purchased $22\frac{5}{8}$ yards from the bolt. How many yards were left on the bolt? _____

5. Vincent lives $10\frac{1}{8}$ miles from his job. His car broke down $4\frac{3}{8}$ miles from his home. How far is he from work? _____

6. Donna started a trip with $18\frac{1}{2}$ gallons of gasoline in her tank. She used $17\frac{3}{4}$ gallons. She bought 14 gallons more. How many gallons does she now have in her tank? _____

7. Kay has a cup of water that measures $\frac{8}{8}$ when full. She poured $\frac{3}{8}$ of it out. How much does she have left? _____

8. Melvin lives $2\frac{11}{16}$ miles from town. If he walked $1\frac{3}{16}$ miles toward town and then got a ride the rest of the way, how far did he ride? _____

Posttest

Add or subtract. Reduce answers to lowest terms.

1. $\begin{array}{r} \frac{5}{16} \\ + \frac{7}{16} \\ \hline \end{array}$

2. $\begin{array}{r} \frac{7}{8} \\ - \frac{1}{8} \\ \hline \end{array}$

3. $\begin{array}{r} \frac{5}{6} \\ \frac{1}{2} \\ + \frac{3}{4} \\ \hline \end{array}$

4. $\begin{array}{r} 7 \\ - \frac{4}{7} \\ \hline \end{array}$

5. $\begin{array}{r} 6\frac{3}{4} \\ - 1\frac{5}{6} \\ \hline \end{array}$

6. $\begin{array}{r} 8\frac{3}{5} \\ + 3\frac{1}{2} \\ \hline \end{array}$

7. $\begin{array}{r} 22\frac{9}{10} \\ - 16\frac{1}{5} \\ \hline \end{array}$

8. $\begin{array}{r} 3\frac{1}{6} \\ - 2\frac{1}{2} \\ \hline \end{array}$

9. $\begin{array}{r} 8\frac{1}{10} \\ - 3\frac{3}{5} \\ \hline \end{array}$

10. $\begin{array}{r} 9\frac{7}{8} \\ - 3\frac{3}{16} \\ \hline \end{array}$

11. $\begin{array}{r} \frac{11}{12} \\ - \frac{1}{2} \\ \hline \end{array}$

12. $\begin{array}{r} 8\frac{1}{6} \\ - 2\frac{5}{6} \\ \hline \end{array}$

Solve the following problems.

13. Ed put whipped butter in a 5-quart pail. Then he filled the pail by adding $3\frac{1}{3}$ qts. of milk. How many quarts are butter?

14. A clerk had a roll of tubing 53 yards long. He sold 5 pieces each $2\frac{1}{2}$ yards long. After the sales, how many yards were left on the roll?

_____

_____

15. The dry ingredients for a cake called for $2\frac{1}{2}$ cups of flour and $1\frac{3}{4}$ cups of sugar. What was the total amount of dry ingredients?

_____

16. Sandra was planning a party. She bought $4\frac{1}{4}$ pounds of peanuts, $2\frac{1}{2}$ pounds of candy, and 2 pounds of cookies. If 15 people attend the party, what will be the average amount of food for each guest?

_____

Circle the correct answer.

17. Donna's net salary is $710 a month. Last month she spent $\frac{1}{4}$ of it for rent, $\frac{1}{8}$ for clothing, and $\frac{1}{8}$ for food. How much money did she have left?
    - **(1)** $255
    - **(2)** $355
    - **(3)** $455
    - **(4)** $555
    - **(5)** $435

18. Mary has $3\frac{1}{2}$ yards of ribbon to cut into six equal pieces. How long will each piece be?
    - **(1)** 6
    - **(2)** $1\frac{5}{7}$
    - **(3)** $\frac{5}{7}$
    - **(4)** $\frac{7}{12}$
    - **(5)** $\frac{7}{6}$

19. Roxanne is $\frac{1}{8}$ as old as her mother. Her mother is 48 years old. How many years older is her mother?
    - **(1)** 6
    - **(2)** 8
    - **(3)** 12
    - **(4)** 42
    - **(5)** 18

20. A travel agent is computing the cost of airline tickets for a couple. One person flies for $\frac{2}{3}$ of the cost of the full fare. What is the total costs of the tickets if full fare is $150?
    - **(1)** $100
    - **(2)** $150
    - **(3)** $200
    - **(4)** $250
    - **(5)** $50

21. Nancy weighs $105\frac{3}{4}$ pounds and Ruby weighs $99\frac{1}{2}$ pounds. What is the difference in their weight?
    - **(1)** $6\frac{1}{2}$
    - **(2)** $6\frac{1}{3}$
    - **(3)** $6\frac{1}{4}$
    - **(4)** $5\frac{3}{4}$
    - **(5)** $6\frac{3}{4}$

22. Using a power paint roller, a painter can paint a 9×12 room in $\frac{3}{4}$ of an hour. Without the power roller, it takes $2\frac{1}{2}$ hours. How much more time does it take to paint without the power roller?
    - **(1)** $2\frac{1}{2}$
    - **(2)** $1\frac{3}{4}$
    - **(3)** $\frac{1}{12}$
    - **(4)** $1\frac{1}{3}$
    - **(5)** $2\frac{3}{4}$

9

Changing Fractions to Decimals and Decimals to Fractions

Complete the chart below.

	Fraction	Decimal
1.	$8\frac{9}{10}$	_____
2.	_____	4.1
3.	$2\frac{12}{100}$	_____
4.	_____	5.09
5.	$\frac{7}{8}$	_____
6.	_____	.625

Compare the following values by using >, <, or =.

7. $\frac{3}{10}$ _____ .3

8. 1.4 _____ $1\frac{1}{2}$

9. 4.8 _____ $4\frac{3}{4}$

10. $4\frac{1}{2}$ _____ 4.6

11. .2 _____ $\frac{1}{5}$

12. $\frac{35}{10}$ _____ 3.55

Solve the following problems. Express answers in both fractional and decimal values.

13. From a bolt of fabric $18\frac{1}{2}$ yards long, a piece 3.75 yards long and another piece $2\frac{7}{8}$ yards were cut off. How much is left on the bolt?

14. Oscar caught one fish $20\frac{7}{8}$ inches long. The longest fish he caught was 21.25 inches. What is the difference in length between the two fish?

_____   _____

Writing Decimals as Fractions or Mixed Numbers

Decimals are fractions too. Their denominators are 10, 100, 1,000, and so on. If a decimal has a whole number before the decimal point, it can be written as a mixed number.

To write a decimal as a fraction, follow these steps:

Step 1 Read the decimal in words. Use the first number as the numerator. Use the "th" word as the denominator.

Step 2 Reduce to lowest terms.

To write a decimal as a mixed number, follow these steps:

Step 1 Read the decimal in words. The whole number comes before the "and." Write it down.

Step 2 Read the fractional part. Write it down.

Step 3 Reduce to lowest terms.

Example

A. Write .625 as a fraction.

Step 1 Read .625 as "625 thousandths." 625 is the numerator. 1,000 is the denominator.

Step 2 Reduce to lowest terms.

$$.625 = \frac{625}{1,000} = \frac{625 \div 125}{1,000 \div 125} = \frac{5}{8}$$

B. Write 6.5 as a mixed number.

Step 1 Read 6.5 as "6 and 5 tenths." 6 is the whole number. 5 is the numerator. 10 is the denominator.

Step 2 Reduce to lowest terms.

$$6.5 = 6\frac{5}{10} = 6\frac{1}{2}$$

Write as fractions or mixed numbers. Reduce to lowest terms.

1. one tenth _____

2. one hundredth _____

3. eighty-nine thousandths _____

4. thirty-two hundredths _____

5. fifty ten-thousandths _____

6. eight thousandths _____

7. one and three tenths _____

8. 7.2 _____

9. .32 _____

10. .001 _____

11. 14.3 _____

12. 310.9 _____

13. 6.735 _____

14. 29.29 _____

15. .503 _____

16. 65.06 _____

17. 11.25 _____

18. 20.75 _____

19. 118.333 _____

20. .225 _____

Changing Fractions to Decimals

To change a fraction to its equivalent decimal, divide the numerator by the denominator.

Examples

Examples A and B are **terminating decimals** because there are no remainders. Example C is a **repeating decimal** because the remainder of 1 keeps repeating. In such a case, stop dividing after two decimal places and write the remainder as a fraction.

A. Change $\frac{3}{4}$ to a decimal.

$$\begin{array}{r} .75 \\ 4\overline{)3.00} \\ -2\,8 \\ \hline 20 \\ -20 \\ \hline \end{array}$$

Divide 3 by 4.
Place a decimal point after the numerator in the dividend.
Add zeros to the right of the decimal point.
Divide as with decimals.

$$\frac{3}{4} = .75$$

B. Change $1\frac{1}{5}$ to a decimal.

$$\begin{array}{r} 1.2 \\ 5\overline{)6.0} \\ -5 \\ \hline 1\,0 \\ -1\,0 \\ \hline \end{array}$$

First change the mixed number to an improper fraction.

$$1\frac{1}{5} = \frac{1 \times 5}{5} + \frac{1}{5} = \frac{6}{5}$$

Divide 6 by 5.

$$1\frac{1}{5} = 1.2$$

C. Change $\frac{1}{3}$ to a decimal.

$$\begin{array}{r} .33\frac{1}{3} \\ 3\overline{)1.00} \\ -\ 9 \\ \hline 10 \\ -9 \\ \hline 1 \end{array}$$

Divide 1 by 3.

$$\frac{1}{3} = .33\frac{1}{3}$$

Change the fractions and mixed numbers to decimals.

1. $\frac{1}{10}$ _____

2. $\frac{1}{100}$ _____

3. $\frac{1}{5}$ _____

4. $\frac{2}{5}$ _____

5. $\frac{1}{2}$ _____

6. $\frac{3}{5}$ _____

7. $\frac{7}{10}$ _____

8. $\frac{4}{5}$ _____

9. $\frac{9}{10}$ _____

10. $\frac{1}{4}$ _____

11. $\frac{3}{4}$ _____

12. $\frac{1}{3}$ _____

13. $\frac{2}{3}$ _____

14. $\frac{1}{8}$ _____

15. $\frac{3}{8}$ _____

16. $\frac{5}{8}$ _____

17. $\frac{7}{8}$ _____

18. $\frac{1}{6}$ _____

19. $\frac{5}{6}$ _____

20. $\frac{3}{10}$ _____

Solve the following problems. Write the mixed numbers as decimals.

21. Manuel ran in Bay City's marathon race. His time was $2\frac{1}{5}$ hours. Write his time as a decimal.

22. Deryl drove 535 miles in $8\frac{1}{2}$ hours. Write his driving time as a decimal.

23. Mr. Duncan babysat for $5\frac{2}{3}$ hours. Write his babysitting time as a decimal.

24. Vivian's flight home took $3\frac{4}{5}$ hours. Write the flight time as a decimal.

Changing Fractions To Decimals Using a Calculator

You can use your calculator to change a fraction to a decimal.
Divide the numerator by the denominator.

Examples

A. To find the decimal for $\frac{3}{4}$, follow these steps on the calculator:

Step 1 Press **3** .

Step 2 Press **÷** .

Step 3 Press **4** .

$\frac{3}{4}$ ← numerator
$\frac{3}{4}$ ← denominator

$\frac{3}{4} = 3 \div 4$

Step 4 Press **=** .

Step 5 Read the answer. Write it down. $3 \div 4 = 0.75$

B. You can also use the calculator to compare fractions. To compare fractions, follow these steps:

Step 1 Change the fractions to decimals. $\frac{5}{8} = 5 \div 8 = .625$

$\frac{11}{16} = 11 \div 16 = .6875$

Step 2 Compare the decimals. $.625 < .6875$ so

$\frac{5}{8} \quad < \quad \frac{11}{16}$

Practice

Find a decimal for each fraction. Use your calculator.

1. $\frac{3}{10}$ _____   **2.** $\frac{4}{5}$ _____   **3.** $\frac{7}{8}$ _____   **4.** $\frac{1}{16}$ _____

Use the decimal answers from the problems above to compare the fractions. Write >, <, or =.

5. $\frac{3}{10}$ ____ $\frac{7}{8}$ **6.** $\frac{1}{16}$ ____ $\frac{4}{5}$ **7.** $\frac{7}{8}$ ____ $\frac{4}{5}$ **8.** $\frac{3}{10}$ ____ $\frac{1}{16}$

174

Comparing Fraction and Decimal Values

Which is larger, 1.25 a mile or $1\frac{1}{4}$ of a mile? When you want to compare a fraction and a decimal, change both values to either fractions or decimals.

Examples

A. Which is larger, $1\frac{1}{4}$ of a mile or 1.25 of mile?

Change 1.25 to a fraction.

$1.25 = 1\frac{25}{100} = 1\frac{1}{4}$

When changed to a fraction 1.25 is the same as $1\frac{1}{4}$. Therefore, 1.25 miles is equal to $1\frac{1}{4}$ miles.

B. Which is longer, $\frac{1}{2}$ of an inch or .75 of an inch?

Change $\frac{1}{2}$ to a decimal.

$$\frac{1}{2} = \begin{array}{r} .50 \\ 2\overline{)1.00} \\ \underline{1\,0} \\ 0 \\ \underline{0} \end{array} = .50$$

When changed to a decimal, $\frac{1}{2}$ is .50, and .50 is less than .75. Therefore, .50 of an inch is less than .75 of an inch.

Practice

Compare the following values by either changing the decimals to fractions or fractions to decimals, using >, <, or =.

1. $4\frac{4}{25}$ _____ 4.25

2. $11\frac{4}{5}$ _____ 11.75

3. $16\frac{1}{4}$ _____ 16.27

4. $1\frac{1}{6}$ _____ $1.16\frac{2}{3}$

5. $1\frac{12}{25}$ _____ 1.2

6. $54\frac{1}{8}$ _____ 54.125

7. 5.625 _____ $6\frac{3}{8}$

8. 2.80 _____ $2\frac{9}{10}$

9. $\frac{5}{3}$ _____ 1.72

10. 215.25 _____ $215\frac{3}{4}$

11. 3.05 _____ $3\frac{1}{5}$

12. $\frac{4}{3}$ _____ $1.33\frac{1}{3}$

LIFE SKILL

Finding Gross Pay Including Overtime

Bill worked 45 hours in one week. He earns $6.00 per hour. For any work hours over 40, he earns **time and a half** or **overtime** pay. What is his gross pay for a 45-hour week?

Find his regular pay.	Find his time-and-a-half rate.	Find his overtime pay.	Gross pay (Regular pay + overtime pay)

$$\begin{array}{r} \$6.00 \\ \times \quad 40 \\ \hline \$240.00 \end{array}$$

$$6.00 \times 1\frac{1}{2}$$

$$\frac{\overset{3.00}{6.00}}{1} \times \frac{3}{\underset{1}{2}} =$$

$9.00 per hour

$$\begin{array}{r} \$9.00 \\ \times \quad 5 \\ \hline \$45.00 \end{array}$$

$$\begin{array}{r} \$240.00 \\ + \quad 45.00 \\ \hline \$285.00 \end{array}$$

Find the gross pay for each employee.

Employee	Time								Rate	Gross Pay
	M	T	W	Th	F	Total Hours	Reg.	O.T.		
1. Cardenas, J.	8	$9\frac{1}{2}$	8	8	9	$42\frac{1}{2}$	40	$2\frac{1}{2}$	$5.00	
2. Park, Y. E.	8	9	$9\frac{1}{2}$	$8\frac{1}{2}$	10	45	40	5	$5.00	
3. Reynolds, K.	$8\frac{1}{2}$	$8\frac{1}{2}$	$8\frac{1}{2}$	$8\frac{1}{2}$	$8\frac{1}{2}$	$42\frac{1}{2}$	40	$2\frac{1}{2}$	$7.00	
4. Chen, L.	$9\frac{1}{2}$	8	$9\frac{1}{2}$	8	$9\frac{1}{2}$	$44\frac{1}{2}$	40	$4\frac{1}{2}$	$7.00	
5. Wills, M.	8	8	8	8	8	40	40	—	$8.00	
6. Lisons, T.	$9\frac{1}{2}$	$9\frac{1}{2}$	$9\frac{1}{2}$	$9\frac{1}{2}$	$9\frac{1}{2}$	$47\frac{1}{2}$	40	$7\frac{1}{2}$	$8.00	

Order of Operations

When solving a math problem, you must ask yourself, Do I add, subtract, multiply, or divide? Sometimes you may have to do more than one of these operations for the same problem.

To do this, you should follow the order of operations shown below:

Step 1 Calculate everything in the parentheses.
Step 2 Simplify all exponents.
Step 3 Multiply or divide from left to right.
Step 4 Add or subtract from left to right.

Symbols you should know:

Addition	$+$	$3 + 4$	$= 7$
Subtraction	$-$	$4 - 3$	$= 1$
Multiplication	$\times$	3×4	$= 12$
	$\cdot$	$3 \cdot 4$	$= 12$
	$(\ )$	$3\,(4)$	$= 12$
Division	$\div$	$12 \div 4$	$= 3$
	—	$\frac{12}{4}$	$= 3$

Examples

A. Find the value of $6(9) + (9 \times \frac{1}{3}) = $ _____.

Multiply inside the parentheses. $6(9) + (9 \times \frac{1}{3}) = $

Multiply, then add. $\qquad\qquad 54 \ + \quad 3 \ = 57$

B. Find the value of $\frac{15-5}{2} - (2 \times 2) =$ _____.

Multiply inside the parentheses.

$$\frac{15-5}{2} - (2 \times 2) =$$

$$\frac{15-5}{2} - 4 =$$

To divide, the operations above the line must be completed.

$$\frac{10}{2} - 4 =$$

Divide, then subtract.

$$5 - 4 = 1$$

Practice

Find the value of each expression. Show your work.

1. $6 \div 3 \times 3 =$ _____

2. $9 \times 4 \div 2 =$ _____

3. $45 \div 5 + 4 \, (3) =$ _____

4. $(4 + 15) - 10 =$ _____

5. $4(9) \times \frac{15}{5} =$ _____

6. $(12 \div 4) \times 15 =$ _____

7. $\frac{(36 \div 6)}{2} - \frac{12}{4} =$ _____

8. $(5 - 2) \times \frac{18}{9} =$ _____

9. $(22 \times \frac{1}{2}) - 4 + (5 \times \frac{1}{5}) =$ _____

10. $(36 \div \frac{1}{6}) \div 6 =$ _____

Problem Solving—Fractions and Decimals

You have already solved problems with whole number, fractional, or decimal answers. Recall these steps:

Step 1 Read the problem and underline the key words. These words will usually relate to some mathematics reasoning computation.

Step 2 Make a plan to solve the problem. Ask yourself, Should I add, subtract, multiply, divide, round, or compare? You may have to do more than one of these operations for the same problem.

Step 3 Find the solution. Use your math knowledge to find your answer.

Step 4 Check the answer. Ask yourself, Is the answer reasonable? Did you find what you were asked for?

Examples

A. Jeremy is making tacos for lunch. He needs $\frac{1}{2}$ cup of cheddar cheese and $\frac{1}{4}$ cup of mozzerella cheese. How much cheese will he need altogether?

Step 1 Determine how much cheese is needed altogether. The key word is **altogether.**

Step 2 The key word indicates addition. Look carefully at the numbers. They are fractions. You must be sure they have a common denominator before you add.

Step 3 Find the solution.

    a. Find a common denominator for $\frac{1}{2}$ and $\frac{1}{4}$.

Multiples of 2	Multiples of 4
$1 \times 2 = 2$	$1 \times 4 = 4$
$2 \times 2 = 4$	

    The common denominator is 4.

    b. Add the fractions and write the sum in lowest terms.

$$\frac{1}{4} = \frac{1}{4}$$
$$+\frac{1}{2} = +\frac{2}{4}$$
$$\frac{3}{4} \text{ cup of cheeses}$$

Step 4 Check the answer. Does it make sense that $\frac{3}{4}$ feet is the sum of $\frac{1}{2}$ and $\frac{1}{4}$? Yes, the answer is reasonable.

B. Lucas and Carmen own a fruit stand. Lucas sold 35.6 pounds of fruit. Carmen sold 48.7 pounds of fruit. How many pounds did they sell altogether? How much more did Carmen sell than Lucas?

Step 1 Determine two things: the total number of pounds sold and how much more Carmen sold. The key words are **altogether** and **more.**

Step 2 The key words indicate the operations addition and subtraction. You will be working with decimals. Remember to bring down the decimal point.

Step 3 Find the solution.

    a. First add.

$$
\begin{array}{r}
4\ 8\ .\ 7 \text{ pounds} \\
+\ 3\ 5\ .\ 6 \text{ pounds} \\
\hline
8\ 4\ .\ 3 \text{ pounds of fruit sold}
\end{array}
$$

    b. Next, subtract.

$$
\begin{array}{r}
4\ 8\ .\ 7 \text{ pounds} \\
-\ 3\ 5\ .\ 6 \text{ pounds} \\
\hline
1\ 3\ .\ 1 \text{ more pounds sold by Carmen}
\end{array}
$$

Step 4 Check the answers. Do they make sense? Yes.

Practice

Solve the following problems.

1. Mr. Hagen planned a 900-mile trip. If he drove 5.9 hours at an average of 52 miles per hour, how many miles would he have left to drive?   _____

2. A bookshelf is 36 inches long. How many books, $1\frac{3}{4}$ inches each, will fit on the shelf?   _____

3. Travis read 30 pages of a 40-page short story. What portion does he have yet to read?   _____

4. What is the value of $9\frac{3}{4} \div 9.75$?   _____

5. The Village Fruit Market had 32.5 pounds of apples. If $8\frac{3}{4}$ pounds of apples were left at the end of the day, how many pounds were sold?   _____

Complete the chart below.

	Decimal	Fraction
1.	_____	$\frac{3}{10}$
2.	.2	_____
3.	.45	_____
4.	_____	$\frac{375}{1000}$
5.	.6	_____
6.	_____	$\frac{4}{100}$
7.	.008	_____
8.	_____	$\frac{25}{1000}$
9.	_____	$1\frac{5}{10}$
10.	_____	$12\frac{75}{100}$

Compare the following values by using >, <, or =.

11. $\frac{5}{8}$ _____ .63

12. 1.5 _____ $1\frac{4}{5}$

13. $\frac{7}{10}$ _____ .625

14. 2.5 _____ $2\frac{1}{3}$

15. $\frac{8}{10}$ _____ .88

16. 2.6 _____ $2\frac{3}{7}$

17. $\frac{11}{15}$ _____ .6

18. .625 _____ $\frac{7}{12}$

19. .4 _____ $\frac{4}{9}$

20. 1.875 _____ $1\frac{7}{9}$

Solve the following problems. Express answers in both fractional and decimal values.

21. A box of fruit weighs $8\frac{1}{2}$ pounds. The empty box weighs .875 pounds. What is the weight of the fruit?

_____

22. Chiang bought three chickens. One weighed $3\frac{7}{8}$ pounds. The other two each weighed 4.125 pounds each. What was the total weight of the three chickens?

_____

Practice for Mastering Fractions

Multiply.

1. $\frac{3}{4} \times \frac{1}{3}$ _____

2. $\frac{5}{8} \times \frac{3}{7}$ _____

3. $\frac{9}{10} \times \frac{7}{8}$ _____

4. $4\frac{3}{8} \times 2\frac{2}{5}$ _____

5. $1\frac{1}{9} \times \frac{3}{5}$ _____

6. $4\frac{1}{5} \times 1\frac{3}{7}$ _____

7. $4\frac{6}{7} \times 28$ _____

8. $2\frac{9}{10} \times 20$ _____

9. $1\frac{1}{2} \times 18$ _____

10. $\frac{11}{12} \times 5\frac{1}{3}$ _____

11. $5\frac{3}{4} \times 12$ _____

12. $16\frac{3}{8} \times 5\frac{1}{3}$ _____

13. $4\frac{1}{8} \times 2\frac{2}{3}$ _____

14. $3\frac{1}{3} \times 5\frac{1}{4}$ _____

15. $5\frac{1}{6} \times 3\frac{3}{4}$ _____

Divide.

16. $\frac{3}{8} \div \frac{1}{2}$ _____

17. $\frac{3}{5} \div \frac{3}{4}$ _____

18. $\frac{2}{9} \div \frac{2}{3}$ _____

19. $\frac{4}{5} \div 2$ _____

20. $\frac{7}{15} \div 3$ _____

21. $\frac{7}{25} \div 4$ _____

22. $12 \div 1\frac{1}{3}$ _____

23. $20 \div 7\frac{1}{2}$ _____

24. $24 \div 1\frac{7}{8}$ _____

25. $3\frac{3}{4} \div 2\frac{1}{2}$ _____

26. $2\frac{5}{8} \div 3\frac{1}{9}$ _____

27. $5\frac{1}{3} \div 9\frac{1}{3}$ _____

28. $5\frac{5}{8} \div 3\frac{1}{8}$ _____

29. $15\frac{1}{3} \div 9\frac{1}{3}$ _____

30. $4\frac{5}{8} \div 1\frac{1}{8}$ _____

Solve the following problems.

31. Mr. Barrios bought $4\frac{1}{2}$ pounds of candy for $2 a pound. How much did he pay for the candy?

_____

32. A store packs dill spice in packages of $1\frac{3}{4}$ ounces. How many ounces are there in 12 packages?

_____

33. If you lose $2\frac{1}{4}$ pounds a week for 5 weeks, how many pounds will you lose?

_____

34. A grocer put 16 pounds of onions into bags holding $\frac{3}{4}$ pounds each. How many bags did he fill?

_____

35. If each pie serving is $\frac{1}{5}$ of a pie, how many customers can be served from 15 pies?

_____

36. A piece of lumber is $21\frac{1}{2}$ feet long. If it is cut into three equal pieces, how long is each piece?

_____

Add. Write answers in lowest terms.

37. $3\frac{2}{3}$
   $+\ 2\frac{5}{6}$

38. $5\frac{1}{3}$
   $+\ 2\frac{3}{4}$

39. $3\,2\frac{1}{8}$
   $+\ 8\,4\frac{5}{6}$

40. $3\frac{2}{5}$
   $+\ 1\frac{1}{2}$

41. $3\frac{1}{9}$
   $+\ 8\frac{3}{4}$

42. $7\frac{5}{8}$
   $+\ 7\frac{1}{16}$

43. $7\frac{5}{32}$
   $+\ 6\frac{13}{16}$

44. $1\,2\frac{7}{9}$
   $+\ 3\frac{5}{6}$

45. $1\frac{7}{8}$
   $+\ 8\frac{1}{2}$

46. $1\,2\frac{1}{6}$
   $4\frac{3}{4}$
   $+\ 5\frac{7}{8}$

47. $5\frac{1}{2}$
   $6\frac{3}{14}$
   $+\ 3\frac{5}{7}$

48. $1\,1\frac{3}{4}$
   $3\,9\frac{3}{8}$
   $+\ \frac{7}{16}$

Solve the following problems.

49. A real estate company had 3 parcels of land to sell. They were $5\frac{5}{6}$ acres, $4\frac{1}{8}$ acres, and $37\frac{2}{3}$ acres. How many acres were for sale?

50. Lupita bought two chickens. One weighed $2\frac{1}{2}$ pounds and the other $3\frac{1}{4}$ pounds. What was the total weight of the two chickens?

Subtract.

51.
$$3\tfrac{1}{5}$$
$$-\;2\tfrac{1}{3}$$

52.
$$2\tfrac{3}{8}$$
$$-\;1\tfrac{3}{16}$$

53.
$$5\tfrac{2}{3}$$
$$-\;2\tfrac{5}{12}$$

54.
$$5\tfrac{7}{9}$$
$$-\;2\tfrac{1}{3}$$

55.
$$1\,1\tfrac{7}{9}$$
$$-\;7\tfrac{5}{6}$$

56.
$$5\tfrac{1}{2}$$
$$-\;2\tfrac{3}{4}$$

57.
$$5\tfrac{1}{8}$$
$$-\;4\tfrac{5}{12}$$

58.
$$7$$
$$-\;2\tfrac{2}{5}$$

59.
$$1\,9\tfrac{5}{9}$$
$$-\;1\,6$$

60.
$$8$$
$$-\;\tfrac{2}{3}$$

61.
$$6\tfrac{2}{9}$$
$$-\;5\tfrac{11}{15}$$

62.
$$9\,3\tfrac{2}{5}$$
$$-\;4\,9\tfrac{7}{8}$$

Solve the following problems.

63. A large can holds 5 cups of fruit juice. If you drink $2\tfrac{1}{2}$ cups, how much is left in the can?

64. From a roll of electrical wire $73\tfrac{7}{8}$ yards long, a piece $15\tfrac{1}{2}$ yards was cut. Later, $42\tfrac{2}{3}$ yards were cut. How many yards were left on the roll?

Circle the number of the correct answer.

65. Which statement is true?
    - (1) 3.4 is greater than $3\frac{1}{2}$.
    - (2) 2.6 miles is less than $2\frac{1}{2}$ miles.
    - (3) 14.8 gallons is equal to $14\frac{1}{2}$ gallons.
    - (4) $5\frac{2}{3}$ hours is equal to $5.33\frac{1}{3}$ hours.
    - (5) $1\frac{1}{3}$ cups is greater than 1.25 cups.

66. The least common denominator for $\frac{1}{2}$, $\frac{1}{3}$, $\frac{1}{4}$, and $\frac{1}{6}$ is
    - (1) 12
    - (2) 18
    - (3) 24
    - (4) 48
    - (5) 72

67. Which of the following values is not reduced to the lowest terms?
    - (1) $\frac{4}{32} = \frac{1}{8}$
    - (2) $\frac{18}{24} = \frac{6}{8}$
    - (3) $\frac{7}{28} = \frac{1}{4}$
    - (4) $\frac{12}{16} = \frac{3}{4}$
    - (5) $\frac{8}{16} = \frac{1}{2}$

68. How many bags of apples each containing 1.25 pounds can be filled from $50\frac{1}{2}$ pounds of apples?
    - (1) 40.4
    - (2) 40.3
    - (3) 40.2
    - (4) 40.1
    - (5) 40

69. The cost of mailing a package through the postal service cost $\frac{2}{5}$ less than that of a private firm. If the private firm charges $15.75 for a package, what would be the postal service charge?
    - (1) $13.45
    - (2) $12.45
    - (3) $11.45
    - (4) $10.45
    - (5) $9.45

70. There are $6\frac{1}{2}$ grapefruits. How many people can be served grapefruit halves?
    - (1) $13\frac{1}{2}$
    - (2) 13
    - (3) $12\frac{1}{2}$
    - (4) 12
    - (5) $11\frac{1}{2}$

Problems 71 and 72 are related.

71. A car was driven 14,572.5 miles last year, averaging $14\frac{1}{2}$ miles to the gallon. If the average cost per gallon was $1.25, how many gallons of gas were used last year?
    - (1) 1,005
    - (2) 1,006
    - (3) 1,007
    - (4) 1,008
    - (5) 1,009

72. Using the data from problem 71, what was the annual cost for gasoline? (Round to the nearest dollar.)
    - (1) $1,256
    - (2) $1,255
    - (3) $1,254
    - (4) $1,253
    - (5) $1,252

73. An auto assembly plant has 2,400 employees. One fifth of them are on vacation and $.33\frac{1}{3}$ have been laid off. How many are actually active?

(1) 1,400   (2) 1,120

(3) 800   (4) 600

(5) 500

74. Joyce had $\frac{3}{5}$ of $100 at the beginning of the week. If she spent .25 of that amount, how much did she have at the end of the week?

(1) $60   (2) $55

(3) $45   (4) $20

(5) $15

75. Which of the following sets of values is arranged from the greatest to the least?

(1) $\frac{3}{4}$ $\frac{1}{4}$ $.33\frac{1}{3}$ (2) $\frac{1}{5}$ .3 $\frac{1}{2}$

(3) $\frac{2}{5}$ .25 $\frac{1}{5}$ (4) $\frac{3}{8}$ .625 $\frac{6}{8}$

(5) $\frac{3}{4}$ $\frac{2}{3}$ .875

76. It takes $1\frac{1}{3}$ hours to make a round trip from the airport to the downtown area. At this rate, how many trips are possible within 8 hours?

(1) 2   (2) 4

(3) 6   (4) 8

(5) 10

77. Erin earns $20,000 a year. If her taxes represent $\frac{1}{5}$ of her earnings, how much does she have left after taxes?

(1) $16,000   (2) $15,000

(3) $12,000   (4) $10,000

(5) $ 4,000

78. Henry has planned a 1500 calorie a day diet. He ate 400 calories for breakfast, 600 calories for lunch, and the balance for dinner. Represent as a fraction his dinner calories.

(1) $\frac{1}{3}$ (2) $\frac{2}{5}$

(3) $\frac{4}{15}$ (4) $\frac{1}{5}$

(5) $\frac{2}{3}$

79. A shirt presser at the cleaners can press a shirt in .25 of an hour. At the same rate, how many shirts can he press in $6\frac{1}{4}$ hours?

(1) 5   (2) 10

(3) 15   (4) 20

(5) 25

80. Which of the following expressions has the greatest value?

(1) $(\frac{1}{2} + \frac{2}{3}) - \frac{1}{4}$ (2) $\frac{3}{4} \times \frac{2}{3} \times \frac{3}{5}$

(3) $\frac{5}{8} + (\frac{7}{9} \div \frac{7}{9})$ (4) $(2\frac{5}{8} - \frac{3}{4}) \div 4\frac{2}{3}$

(5) $6 \times (\frac{2}{3} + \frac{1}{2})$

Fractions Posttest

Write a fraction, mixed number, or whole number for the shaded portions of the pictures below. Reduce to lowest terms.

1.

2.

3.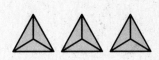

4.

_____   _____   _____   _____

Change the following to a whole number, mixed number, or improper fraction.

5. $\frac{15}{7}$ _____   **6.** $7\frac{2}{3}$ _____   **7.** $\frac{15}{2}$ _____   **8.** $\frac{16}{4}$ _____

Change the following fractions to decimals.

9. $\frac{1}{3}$ _____   **10.** $\frac{1}{5}$ _____   **11.** $2\frac{1}{2}$ _____   **12.** $3\frac{3}{8}$ _____

Change the following decimals to fractions or mixed numbers.

13. .875 _____   **14.** 7.50 _____   **15.** .3 _____   **16.** .6666 _____

Compare the following fractions and decimal values by using >, <, or =.

17. $5\frac{3}{5}$ _____ $5\frac{6}{10}$ **18.** $\frac{2}{3}$ _____ $3\frac{1}{3}$ **19.** 3.375 _____ $3\frac{3}{8}$ **20.** $\frac{5}{8}$ _____ .625

21. .24 _____ $\frac{1}{4}$ **22.** .125 _____ $\frac{1}{8}$ **23.** $12\frac{3}{4}$ _____ 12.8   **24.** 3.6 _____ $3\frac{1}{2}$

Divide. Write the quotients as mixed numbers.

25. $7\overline{)100}$ **26.** $15\overline{)185}$ **27.** $6\overline{)172}$ **28.** $10\overline{)2,236}$

29. Change the fractions to decimals. Then arrange the decimals from the smallest to the largest.

$\frac{14}{25}$ $\frac{8}{3}$ $\frac{13}{20}$ $\frac{4}{5}$ $\frac{5}{8}$

_____

30. Change the decimals to fractions. Then arrange the fractions from the largest to smallest.

.5 .25 .125 .333 .2

_____

Multiply. Reduce to lowest terms.

31. $\frac{4}{7} \times \frac{14}{16}$ _____ **32.** $\frac{1}{4} \times 15$ _____ **33.** $5 \times \frac{7}{20}$ _____ **34.** $2\frac{1}{2} \times 8$ _____

35. $\frac{4}{7} \times 4\frac{1}{5} \times 5$ _____ **36.** $5\frac{1}{4} \times 1\frac{2}{7} \times 1\frac{1}{15}$ _____

Divide. Reduce to the lowest terms.

37. $\frac{5}{6} \div \frac{1}{3}$ _____ **38.** $\frac{25}{36} \div 10$ _____ **39.** $12 \div \frac{3}{5}$ _____ **40.** $3\frac{1}{3} \div 8\frac{1}{3}$ _____

41. $2\frac{1}{4} \div 1\frac{1}{4}$ _____ **42.** $4\frac{1}{8} \div 2\frac{1}{4}$ _____ **43.** $\frac{11}{20} \div 1\frac{5}{6}$ _____ **44.** $3\frac{2}{3} \div \frac{1}{12}$ _____

Add. Reduce to the lowest terms.

45.
$$\begin{array}{r} \frac{2}{5} \\ \frac{1}{5} \\ + \frac{6}{10} \\ \hline \end{array}$$

46.
$$\begin{array}{r} 2\frac{1}{2} \\ 7\frac{1}{4} \\ + 5\frac{1}{5} \\ \hline \end{array}$$

47.
$$\begin{array}{r} 3\frac{2}{3} \\ 7\frac{1}{6} \\ + 2\frac{1}{4} \\ \hline \end{array}$$

48.
$$\begin{array}{r} 1\frac{3}{8} \\ \frac{1}{3} \\ + 2\frac{1}{4} \\ \hline \end{array}$$

49.
$$\begin{array}{r} 11\frac{1}{8} \\ + 2\frac{5}{12} \\ \hline \end{array}$$

50.
$$\begin{array}{r} 9\frac{3}{14} \\ + \frac{2}{3} \\ \hline \end{array}$$

51.
$$\begin{array}{r} 17\frac{1}{5} \\ + 41\frac{3}{4} \\ \hline \end{array}$$

52.
$$\begin{array}{r} 7\frac{3}{11} \\ + 5\frac{1}{4} \\ \hline \end{array}$$

Subtract. Reduce to the lowest terms.

53.
$$\begin{array}{r} 88\frac{5}{6} \\ - 53\frac{1}{6} \\ \hline \end{array}$$

54.
$$\begin{array}{r} 13\frac{1}{12} \\ - 7\frac{11}{12} \\ \hline \end{array}$$

55.
$$\begin{array}{r} 10\frac{1}{2} \\ - 3\frac{2}{7} \\ \hline \end{array}$$

56.
$$\begin{array}{r} 12\frac{3}{5} \\ - 9\frac{5}{6} \\ \hline \end{array}$$

57.
$$\begin{array}{r} 30\frac{4}{5} \\ - 15\frac{1}{2} \\ \hline \end{array}$$

58.
$$\begin{array}{r} 5 \\ - \frac{2}{3} \\ \hline \end{array}$$

59.
$$\begin{array}{r} 4\frac{1}{2} \\ - 1\frac{5}{6} \\ \hline \end{array}$$

60.
$$\begin{array}{r} 7\frac{1}{6} \\ - \frac{5}{6} \\ \hline \end{array}$$

Compare the following expressions, using >, <, or =.

61. $1 - \frac{3}{4}$ _____ $\frac{1}{8} + \frac{1}{8}$

62. $\frac{7}{8} \div \frac{1}{2}$ _____ $2\frac{1}{2} + 1\frac{1}{4}$

63. $\frac{1}{2} + \frac{5}{6}$ _____ $\frac{1}{8} + \frac{1}{3}$

64. $3 \div \frac{1}{2}$ _____ $3 + \frac{1}{2}$

Solve the following problems. Circle the correct answer.

65. Which of the following pairs is not equivalent?

   (1) $\frac{20}{25} = \frac{3}{5}$ **(2)** $\frac{4}{5} = \frac{8}{10}$

   (3) $\frac{1}{2} = \frac{16}{32}$ **(4)** $\frac{4}{6} = \frac{6}{9}$

   (5) $\frac{3}{4} = \frac{21}{28}$

66. Find the value of $(3\frac{1}{3} + 2\frac{5}{6}) \div (7\frac{2}{3} - 5\frac{1}{5})$.

   (1) $6\frac{1}{6}$ **(2)** 6

   (3) $2\frac{1}{2}$ **(4)** $\frac{1}{5}$

   (5) $\frac{1}{6}$

67. Millicent bought 7.75 yards of fabric. Alice bought $5\frac{7}{8}$ yards of the same fabric. What was the combined yardage of the two purchases?

   (1) 13.75 **(2)** $13\frac{7}{8}$

   (3) $13\frac{5}{8}$ **(4)** 13.5

   (5) $13\frac{1}{4}$

68. Find the length of the longest side in the diagram below.

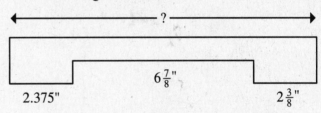

   (1) 11.875 **(2)** 11.625

   (3) 11.5 **(4)** 11.25

   (5) none of these

69.

Doug's Budget

Rent	$300
Food	$250
Child Care	$200
Savings	$100
Utilities	$250

If Doug earns $1,500 a month, what fractional part of his salary goes for childcare and savings?

   (1) $\frac{7}{15}$ **(2)** $\frac{1}{3}$

   (3) $\frac{1}{5}$ **(4)** $\frac{1}{6}$

   (5) $\frac{1}{15}$

70. The purchase price of a compact disc player is $150. The same item rents for $5 per week. Express as a mixed number the rental price versus the purchase price. (*Hint:* The numerator will be the rental price based on 52 weeks a year.)

   (1) $1\frac{1}{2}$ **(2)** $1\frac{3}{5}$

   (3) $1\frac{11}{15}$ **(4)** $\frac{26}{15}$

   (5) $\frac{15}{26}$

71. Mrs. Ohira had $7\frac{1}{2}$ yards of material to make aprons for the bazaar. She needs $1\frac{1}{2}$ yards of material for each apron. How many aprons can she make?

(1) 10   (2) 5

(3) 6   (4) 4

(5) 7

72. Leah planted $1\frac{1}{4}$-foot high tomato plants a month ago. Now they are $2\frac{1}{3}$ feet high. How much did the plants grow in a month?

(1) $1\frac{11}{12}$ ft.   (2) $1\frac{1}{12}$ ft.

(3) $2\frac{1}{6}$ ft.   (4) 2 ft.

(5) $2\frac{1}{12}$ ft.

73. Yoram worked $8\frac{1}{2}$ hours on a project that should take 15 hours to complete. How many more hours does he have to work?

(1) $7\frac{1}{2}$ hr.   (2) 6 hr.

(3) 7 hr.   (4) $6\frac{1}{2}$ hr.

(5) $7\frac{3}{4}$ hr.

74. If Carrie rides $3\frac{1}{4}$ miles round trip on her bike to school in one week, how many miles does she ride in two weeks?

(1) $23\frac{1}{2}$ mi.   (2) $16\frac{1}{4}$ mi.

(3) $6\frac{1}{2}$ mi.   (4) $32\frac{1}{2}$ mi.

(5) none of these

75. Benson bought .5 pounds of nails, $\frac{3}{4}$ pounds of wood screws, 1.25 pounds of quarter inch bolts, and $\frac{3}{4}$ pounds of toggle bolts. What was the total weight of his purchase?

(1) 3.0 lbs   (2) $3\frac{1}{4}$ lbs

(3) 3.5 lbs   (4) $3\frac{3}{4}$ lbs

(5) $-3\frac{1}{4}$ lbs

76. A trip from Denver to Grand Junction takes $5\frac{1}{2}$ hours by car or 1.25 hours by plane. How many hours can be saved by taking the plane?

(1) 4.5 hrs.   (2) 4.375 hrs.

(3) 4.25 hrs.   (4) 4.125 hrs.

(5) 4.325 hrs.

77. Find the length of the shaded portion in the diagram below.

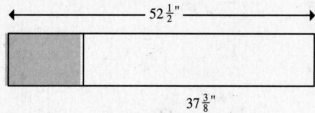

(1) 15.5 in.   (2) 15.25 in.

(3) 15.125 in.   (4) 15.0 in.

(5) 15.155

78. A closet has 36 hangers. Twelve were empty. What fractional number of hangers had clothes?

(1) $\frac{2}{3}$ (2) $\frac{3}{8}$

(3) $\frac{1}{3}$ (4) $\frac{1}{4}$

(5) $\frac{3}{4}$

Pretest Unit/pages 1–3

1. 14
2. 42
3. 5
4. 14
5. 7
6. 7
7. tens
8. ones
9. thousands
10. thousands
11. millions
12. hundred thousands
13. >
14. <
15. >
16. <
17. <
18. <
19. <
20. >
21. >
22. 5,670
23. 76,000
24. 360,000
25. 300,000
26. 3,000; 2,410; 2,114; 2,069
27. 1,551; 1,536; 1,531; 1,510
28. 63
29. 148
30. 356
31. 6 R1
32. 252
33. 7,799
34. 233 R58
35. 432
36. 104,284
37. 354 R118
38. 1,341,490
39. 1,527,106
40. 4,731,317
41. 805,533
42. 1,000 (est), 1,001
43. 4,100 (est), 4,152
44. 85,800 (est), 84,693
45. 20 (est), 21 R12
46. (4) 9
47. (2) 55
48. (2) $85
49. (1) 146
50. (3) 187
51. (1) 42

Lesson 1/page 4

1. 12
2. 12
3. 14
4. 6
5. 26
6. 35
7. 81
8. 65
9. 28
10. 40
11. 12
12. 0
13. $(8 \times 5) + 3 = 43$
14. $(15 - 9) \times 9 = 54$
15. $(4 \times 2) + 5 = 13$
16. $4 + (2 \times 3) = 10$
17. $3 + (5 \times 4) = 23$
18. $18 \div (6 + 3) = 2$
19. $(4 + 2) \times 3 = 18$
20. $(3 + 5) \times 4 = 32$
21. $(18 \div 6) + 3 = 6$

Lesson 2/page 5

1. ones
2. hundreds
3. ones
4. hundreds
5. ones
6. tens
7. hundreds
8. tens
9. 6,987
10. 9,856
11. 7,698
12. 9,856

Lesson 3/page 6

1. ten thousands
2. ten thousands
3. millions
4. hundred thousands
5. hundred thousands
6. thousands
7. thousands
8. ten thousands
9. 8,245,064
10. 53,704
11. 1,521,251
12. 8,245,064
13. 54,276
14. 86,304
15. 27,000
16. 367,913

Lesson 4/pages 7-8

1. <
2. >
3. <
4. >
5. <
6. >
7. >
8. >
9. <
10. >
11. >
12. >
13. >
14. <
15. <
16. >
17. <
18. =
19. >
20. >
21. >
22. >
23. <
24. >
25. <
26. >
27. 86, 76, 74, 69
28. 121, 119, 116, 106
29. 2,912; 2,335; 2,291; 2,264
30. 700, 597, 498, 488
31. 488,700; 408,587; 48,900
32. 189,501; 110,999; 103,273
33. 317, 357, 371, 383
34. 8,648; 8,686; 8,863; 8,888
35. 1,707; 1,717; 1,770; 1,771
36. 512,675; 512,876; 521,675

Lesson 5/page 9

	tens	hundreds	thousands
1.	8,440	8,400	8,000
2.	7,690	7,700	8,000
3.	3,660	3,700	4,000

	ten thousands	hundred thousands
4.	750,000	800,000
5.	330,000	300,000
6.	170,000	200,000

Lesson 6/pages 10-11

1. 258
2. 349
3. 230
4. 288
5. 36
6. 3
7. 293
8. 430
9. 173
10. 88
11. 141
12. 94
13. 1,214
14. 1,441
15. 1,515
16. 1,661
17. 10,164
18. 15,139
19. 9,185
20. 12,054
21. 10,936
22. 9,321
23. 18,087
24. 13,739
25. 571
26. 430
27. 166
28. 475
29. 6,878
30. 8,744
31. 5,177
32. 6,797
33. 1,932
34. 1,830
35. 2,687
36. 33
37. 37,663
38. 29,396
39. 2,515
40. 734

Lesson 7/pages 12-14

1. 227 trees
2. 1,736 gallons
3. 12,919 seats
4. $97
5. 527 people
6. 14 invitations
7. 1,775 cars
8. 21,619 votes
9. 10,275 hamburgers
10. 105 sets

Lesson 8/pages 15-17

1. 442
2. 3,816
3. 1,664
4. 1,508
5. 48,852
6. 17,675
7. 38,250
8. 47,151
9. 13,338
10. 8,756
11. 11,700
12. 35,295
13. 66,420
14. 135,336
15. 74,617
16. 56,098
17. 93 R2
18. 823
19. 853
20. 489
21. 78
22. 87 R1
23. 89
24. 76
25. 38
26. 144
27. 134 R3
28. 206
29. $12
30. $31,306
31. $88
32. $60
33. 120 items
34. 296 seats
35. 184 people

Posttest Unit 1/pages 18-20

1. 11
2. 72
3. 3
4. 14
5. 6
6. 36
7. tens
8. ones
9. thousands
10. hundreds
11. hundred thousands
12. hundred thousands
13. <
14. <
15. >
16. >
17. >
18. >
19. >
20. >
21. <
22. 7,400
23. 17,000
24. 6,000
25. 330,000
26. 800,000
27. 54,500

28. 9,587; 9,475; 9,209; 8,990
29. 5,987; 5,976; 5,971; 5,967
30. 29,110; 29,035; 28,990; 28,799
31. 385,800; 379,900; 374,900
32. 7,545
33. 19,086
34. 86,801
35. 3 R4
36. 231 R41
37. 19,593
38. 126,950
39. 920,979
40. 2,129,831
41. 347 R25
42. 942,037
43. 3,452
44. 2,228 R82
45. 1,839,809
46. 9,000 (est), 9,036
47. 3,000 (est), 2,822
48. 300,000 (est), 286,792
49. 250 (est), 236
50. (4)
51. (4)
52. (4)
53. (3)
54. (2)
55. (4)
56. (5)
57. (2)

Pretest Unit 2/page 21

1. thousandths
2. tenths
3. 9.5
4. $6.88
5. 0.081
6. 0.8
7. 127.56
8. .085
9. 0.0009
10. 10.6 1.06 .106
11. 8.4 8.04 .804

Lesson 9/page 22

1. $6.90
2. $15.03
3. $0.09
4. $0.10
5. 7 cents
6. 7 dollars
7. 700 dollars
8. 70 cents

Life Skill/page 23

1. $56.40
2. $29.11
3. $18.09
4. $299.95
5. $101.00
6. $30.97

Lesson 10/pages 24-26

The exercises on page 24 are oral.
1. thousandths
2. ten-thousandths

3. hundred-thousandths
4. millionths
5. 57.4
6. 0.0312
7. 0.70
8. 5,500.05
9. 53.000053
10. 7.00007
11. fifteen and eighteen hundredths
12. thirteen and thirteen hundred-thousandths
13. fifteen thousandths
14. five and ninety-seven millionths
15. seven tenths
16. two ten-thousandths
17. h
18. a
19. j
20. g
21. f
22. e
23. d
24. b
25. i
26. c
27. hundredths
28. tenths
29. thousandths
30. ten-thousandths
31. tenths
32. ten-thousandths
33. thousandths
34. hundred-thousandths
35. millionths
36. hundredths

Lesson 11/pages 27-28

1. 5, 6, 7
2. 6.1; 6.2; 6.4; 6.6; 6.8; 6.9
3. 0.72; 0.73; 0.76; 0.78; 0.79
4. 2.200; 2.300; 2.400; 2.600; 2.800; 2.900
5. G
6. J
7. O
8. C
9. A
10. E
11. R
12. K
13. Q
14. H
15. M
16. N
17. I
18. D
19. F
20. P

Lesson 12/pages 29-30

1.	>	20.	<
2.	>	21.	<
3.	>	22.	<
4.	>	23.	<
5.	>	24.	<
6.	>	25.	>
7.	>	26.	=
8.	<	27.	<
9.	=	28.	=
10.	<	29.	<
11.	<	30.	>
12.	>	31.	=
13.	<	32.	<
14.	>	33.	>
15.	=	34.	<
16.	>	35.	b
17.	<	36.	e
18.	<	37.	b
19.	<	38.	e

Lesson 13/page 31

1. .02 .15 .2 11.015
2. .023 .032 .23 .32
3. 3.75 .375 .0375 .00375
4. .125 .1025 .025 .0125

Life Skill/pages 32-33

Answers can be found on page 202.

Lesson 14/pages 34-35

1. 0.1 0.05
2. 17.5 17.52
3. 785.4 785.43
4. 79.4 79.37
5. 0.7 0.74
6. 66.2 66.18
7. 99.9 99.88
8. 137
9. 136
10. 7
11. 6
12. 18
13. 19
14. 0.4
15. 0.3
16. 519.4
17. 43,500.2
18. 0.7
19. 0.7
20. 276.96
21. 276.96
22. 68.11
23. 15.83
24. 45.13
25. 86.58
26. 0.108
27. 2.213

28. .095
29. 58.2360
30. 0.0229
31. 13.9091
32. 10.87533
33. 0.50057
34. 429.30477

Life Skill/page 36

1.	$2.00	7.	.60
2.	.50	8.	1.50
3.	.30	9.	1.10
4.	1.80	10.	.80
5.	.70	11.	.50
6.	.70	12.	.50

exact total $11.03
estimate total $11.00
difference $.03

Lesson 15/pages 37-38

1. Ms. Kasey
2. 6 shares
3. 88
4. greatest: Denise; least: Henry
5. yes

Posttest Unit 2/page 39

1. $4.59
2. $0.02
3. $0.70
4. $1.05
5. $3.03
6. thousandths
7. tenths
8. ones
9. hundredths
10. hundred-thousandths
11. .02 and .020
12. 0.19 and .190
13. .3 and .3000
14. 6.66 and 6.660
15. <
16. >
17. <
18. <
19. 8.32 8.032 7.3
20. .05 .005 .0005
21. .505 .0505 .0025
22. 1,141.7
23. 1,141.547
24. 1,141.5178

Pretest Unit 3/page 40

1.	134.674	5.	164.9647
2.	47.3042	6.	24.00518
3.	23.49151	7.	$38.05
4.	22.625	8.	0.981

9. $4.79
10. 0.0222
11. .394
12. $7.35
13. 4.129
14. 1.95 feet
15. 39.92 miles

Lesson 16/pages 41-42

1.	2.3	10.	172.103
2.	22.7	11.	3809.853
3.	135.657	12.	$179.33
4.	414.783	13.	$3.53
5.	748.24	14.	33.4 gallons
6.	293.565	15.	$69.24
7.	94.7353	16.	$213
8.	152.235	17.	$142.65
9.	1.13475		

Life Skill/page 43

1. $25.93
2. $18.97

Lesson 17/pages 44-46

1.	$144.76	20.	964.0239
2.	8.886	21.	2.599
3.	1.815	22.	6.2211
4.	$719.76	23.	.01249
5.	0.15	24.	6.449
6.	$194.89	25.	3.104
7.	0.1999	26.	$448.33
8.	$117.35	27.	1.3811
9.	0.4318	28.	607.2655
10.	5.50601	29.	.0069
11.	623.222	30.	$182.59
12.	0.505	31.	165.8 miles
13.	1,196.241	32.	$35.79
14.	65.27	33.	0.5 seconds
15.	.0947	34.	$12.61
16.	287.1503	35.	7.6 feet
17.	.008	36.	$35.15
18.	$200.05	37.	$42.50
19.	17.4405	38.	$73.60

Life Skill/page 47

1. B
2. B
3. $8.22
   2 pennies, 2 dimes
   3 one-dollar bills
   1 five-dollar bill

Lesson 18/pages 48-49

1. 3,143.6 miles
2. $67.50
3. $2.55 million
4. $1.25
5. $6.60

Life Skill/pages 50-51

Answers 1-3 are on page 202.
1. correct
2. incorrect; should be $903.91
3. incorrect; should be $159.26
4. correct
5. correct
6. incorrect; should be $4,499.38
7. incorrect; should be $348.56
8. correct

Lesson 19/pages 52-53

1. 30.4208
2. 179.8685
3. 2.45312
4. 349.143
5. 19.98
6. .19122

Life Skill/page 54

Empire total: $13,221.53
Bluebird total: $11,826
Purchased the Bluebird

Posttest Unit 3/page 55

1. 248.722
2. 508.998
3. 10.0871
4. 18.5908
5. 38.94933
6. 4.63
7. $6.39
8. 6.8136
9. 561.11
10. 13.2643
11. $13.62
12. 8.923 inches
13. 1.2 miles
14. $25.98
15. 1.15 pounds

Pretest Unit 4/pages 56-57

1. 1.6
2. 1,475.5
3. 0.0024
4. 0.0225
5. 0.89241
6. 5.076
7. 46.07
8. 460.7
9. 4,607
10. 1.1
11. 0.695
12. 1.24
13. 0.07
14. .16
15. 217
16. 46.07
17. 4.607
18. 0.4607
19. $2.95
20. $4.00
21. $109.50
22. $28,408.50

Lesson 20/page 58

1. 3
2. 6
3. 2
4. 1
5. 0
6. 4
7. 169.71 16.971 1.6971
8. 6,983.05 698.305 69.8305
9. 541.73 54.173 5.4173
10. 189.63 18.963 1.8963
11. 4,183.27 418.327 41.8327

Lesson 21/pages 59-61

1. 497.35
2. 0.05184
3. 44.25
4. 278.4
5. 399.383
6. 2,244.17
7. 215.636
8. 0.40342
9. 0.016
10. .01
11. .0096
12. 0.00365
13. 0.006936
14. .0039375
15. .0609336
16. .0036363
17. 300.30 miles
18. 6.6804 inches
19. $13.86
20. $54.78
21. 750 blocks, round trip
22. $378
23. $8.30
24. $3.80
25. 335.3 miles

Life Skill/pages 62-64

1. $339
2. $43.93; $100 and $16
3. $159.93
4. $179.07
5. $279.38
6. $2.45; $58.36 and $39.50
7. $100.31
8. $179.07

Check Stubs:

1. $177.52
2. $282.17
3. $158.46
4. $785.71
5. $359.52
6. $256.26

Lesson 22/page 65

1. 76
2. 1.3
3. 76.4
4. 0.13
5. 7.839
6. 0.2
7. 78.39
8. 9.084
9. 783.9
10. 908,400
11. 79
12. 9,084

Life Skill/page 66

1. $67.04
2. $27.80
3. $9.87
4. $21.55
5. $7.08
6. $3.48
7. $3.18
8. $140.00
9. $69.90
10. $17.98
11. $44.75
12. $47.70
13. $12.99
14. $193.32

Lesson 23/page 67

1. 8.28
2. .316
3. 3.759
4. .253
5. 8.9
6. 4.5
7. 45.2
8. 45.5

Life Skill/page 68

1. *Home*
   Total: $17.82
   Per person: $2.97

2. *Restaurant*
   Roast beef: $39
   Milk: $5.10
   Cake: $7.50
   Total: $60.78
   Per person: $10.13
3. *Pizza*
   Total: $12.47
   Per person: $2.08
4. *Fast Food*
   Hamburgers: $5.70
   Fries: $2.70
   Soft drinks: $2.10
   Total: $11.66
   Per person: $1.94
   Most expensive: roast beef
   Least expensive: fast food

Lesson 24/pages 69-70

1. 83.1
2. 1.6
3. .15
4. 45
5. 82
6. .61
7. 1.03
8. 6.22
9. 662
10. 11 books
11. 11.06 inches
12. 14 packages
13. 45 lots

Life Skill/page 71

Answers are on page 202.

Lesson 25/page 72

1. 40.6
2. .0205
3. 202
4. .0493

Lesson 26/page 73

1. .77
2. .014
3. .08306
4. .529
5. .0647
6. .05073

Life Skill/pages 74-76

1. $346.53
2. $270.00
3. $475.00

Payroll Register/page 76

	Gross	Total	Net
Amos	$344.00	$79.34	$264.66
Bryant	$270.00	$86.32	$183.68
Sanchez	$475.00	$80.16	$394.84
Wilkins	$346.53	$39.67	$306.86

Lesson 27/pages 77-78

1. 3.0
2. 0.4
3. 6.5
4. 5.3
5. 2.7
6. 8.6
7. 32.8
8. 7.4
9. 1.45
10. 2.44
11. 1.99
12. 1.49
13. .54
14. .05

Life Skill/page 79

1. Tax: $.47, Total: $9.94
2. Tax: $1.17, Total: $24.52
3. Tax: $1.86, Total: $39.14

Lesson 28/pages 80-81

1. 120
2. 72
3. 6,000
4. 21,000
5. 80
6. 4,800
7. 5
8. 2
9. 3
10. 2,150 (Answers may vary.)
11. 22 (Answers may vary.)
12. 30 (Answers may vary.)

Lesson 29/pages 82-84

1. $48
2. 9 sections
3. (1)
4. (3)
5. (5)
6. (4)
7. (4)

Lesson 30/page 85

1. .02
2. 9.82
3. 31.35

Posttest Unit 4/pages 86-87

1. .0122
2. 6.16
3. .05652
4. 4.087
5. 2.715
6. 27.15
7. .006
8. 11.1
9. 16.32
10. 24.7
11. 0.29
12. 0.272
13. 400
14. 20
15. 3
16. $15.67
17. 270.1 gallons
18. $62.10
19. $512.50

Practice for Mastering Decimals/pages 88-90

1. c
2. e
3. b
4. a
5. d
6. <
7. <
8. =
9. >
10. <
11. <
12. .035 .35 3.5 35.1
13. .003 .030 .035 3.33
14. .08 .088 .8 .808
15. 26.415
16. 46.0361
17. 1,324.48
18. 13.57
19. 1,375
20. 137.5
21. 1,248.955
22. 691.207
23. 12.527
24. $1,652.40
25. .3515
26. .0003
27. 3.264
28. 1.03
29. 1.87
30. 98.006
31. 9.8006
32. .98006
33. 80.8
34. 1.80
35. (5)
36. (4)
37. (4)
38. (2)
39. (2)
40. (3)
41. (3)
42. (1)

Decimals Posttest/page 91-92

1. 8.4
2. .000048
3. 840
4. 48,000,000
5. >
6. <
7. <
8. =
9. 65.1; 65; 6.5; .065
10. 730.0; 73; 07.3; .0079
11. I
12. C
13. F
14. A
15. 503.99
16. 1,390
17. 107.4826
18. 5,485
19. $6.85
20. 419,730
21. 316,130
22. 316.13
23. 12.5
24. 17
25. 1.3
26. .129
27. 13.7
28. 66.08
29. 3.420
30. (4)
31. (1)
32. (1)
33. (4)

Pretest Unit 5/page 93

1. $\frac{2}{5}$
2. $\frac{3}{3} = 1$
3. no
4. yes
5. $\frac{3}{8}$
6. $\frac{1}{2}$
7. 72
8. 8
9. 12
10. 30
11. >
12. =
13. >
14. =
15. 18
16. 3
17. 10
18. 29

Lesson 31/pages 94-95

1. 13
2. 9
3. 4
4. 93
5. 18
6. 46
7. 74
8. 50
9. 48
10. 28
11. 27
12. 2
13. 0
14. 76
15. 10
16. 22
17. <
18. >
19. >
20. <
21. >
22. =
23. <
24. >

Lesson 32/pages 96-98

1. 5, fifths
2. 10, tenths
3. 16, sixteenths
4. 6, sixths
5. 8, eighths
6. 12, twelfths
7. $\frac{1}{3}$
8. $\frac{5}{6}$
9. $\frac{3}{4}$
10. $\frac{4}{7}$
11. $\frac{5}{8}$
12. $\frac{1}{2}$
13. 8; 3; $\frac{3}{8}$
14. 12; 5; $\frac{5}{12}$
15. $\frac{5}{36}$
16. $\frac{50}{60} = \frac{5}{6}$
17. $\frac{8}{24} = \frac{1}{3}$
18. $\frac{70}{320} = \frac{7}{32}$
19. $\frac{3}{12} = \frac{1}{4}$

Lesson 33/pages 99-101

1. $\frac{1}{2} = \frac{2}{4}$
2. $\frac{1}{2} = \frac{3}{6}$
3. $\frac{1}{5} = \frac{2}{10}$
4. $\frac{6}{18} = \frac{1}{3}$
5. $\frac{6}{10} = \frac{3}{5}$
6. $\frac{16}{16} = \frac{1}{1}$
7. $\frac{4}{9}$
8. $\frac{4}{60}$
9. $\frac{4}{5}$
10. $\frac{40}{100}$
11. $\frac{4}{7}$
12. $\frac{1}{3}$
13. $\frac{15}{25}$
14. $\frac{6}{7}$
15. $\frac{24}{32}$
16. $\frac{5}{12}$
17. $\frac{5}{9}$
18. $\frac{14}{30}$
19. $\frac{6}{54}$
20. $\frac{18}{40}$
21. $\frac{2}{4}$
22. $\frac{4}{8}$
23. $\frac{8}{16}$
24. $\frac{2}{8}$
25. $\frac{4}{16}$
26. $\frac{2}{16}$
27. $\frac{6}{8}$
28. $\frac{12}{16}$
29. yes
30. yes
31. no
32. no
33. no
34. yes
35. yes
36. yes

Lesson 34/pages 102-103

1. $\frac{1}{3}$
2. $\frac{1}{6}$
3. $\frac{5}{7}$
4. $\frac{2}{3}$
5. $\frac{1}{8}$
6. $\frac{8}{9}$
7. $\frac{3}{5}$
8. $\frac{4}{5}$

9. $\frac{3}{10}$
10. $\frac{4}{25}$
11. $\frac{3}{4}$
12. $\frac{1}{2}$
13. $\frac{7}{9}$
14. $\frac{2}{9}$
15. $\frac{6}{7}$
16. $\frac{4}{9}$
17. $\frac{3}{4}$
18. $\frac{2}{5}$
19. $\frac{1}{3}$
20. $\frac{7}{9}$
21. $\frac{10}{13}$
22. $\frac{7}{8}$
23. $\frac{1}{4}$
24. $\frac{2}{3}$
25. $\frac{1}{4}$
26. $\frac{2}{5}$
27. $\frac{1}{9}$
28. $\frac{2}{3}$
29. $\frac{2}{7}$
30. $\frac{3}{10}$
31. $\frac{7}{8}$
32. $\frac{1}{2}$
33. $\frac{1}{2}$
34. $\frac{4}{5}$

35. $\frac{4}{9}$
36. $\frac{1}{8}$
37. $\frac{5}{8}$
38. $\frac{5}{6}$
39. $\frac{5}{7}$
40. $\frac{1}{9}$
41. $\frac{3}{8}$
42. $\frac{2}{3}$
43. $\frac{2}{3}$
44. $\frac{1}{9}$
45. $\frac{2}{3}$
46. $\frac{5}{11}$
47. $\frac{1}{5}$
48. $\frac{4}{7}$
49. $\frac{1}{6}$
50. $\frac{1}{3}$
51. $\frac{1}{6}$
52. $\frac{5}{6}$
53. $\frac{1}{3}$
54. $\frac{10}{11}$
55. $\frac{4}{5}$
56. $\frac{2}{3}$
57. (2)
58. (2)
59. (1)
60. (2)
61. (3)

Life Skill/page 104

1. $\frac{1}{4}$
2. $\frac{1}{3}$
3. $\frac{1}{2}$
4. $\frac{2}{3}$
5. $\frac{3}{4}$
6. $\frac{1}{5}$
7. $\frac{1}{4}$
8. $\frac{1}{2}$
9. $\frac{3}{4}$
10. $\frac{4}{5}$
11. $\frac{1}{4}$
12. $\frac{1}{3}$
13. $\frac{1}{2}$
14. $\frac{2}{3}$
15. $\frac{3}{4}$

Lesson 35/pages 105-107

1. 10
2. 8
3. 12
4. 15
5. 15
6. 34
7. 24
8. 33
9. 30
10. 80
11. 26
12. 35
13. 30
14. 24
15. 12
16. 6
17. 15
18. 20
19. 56
20. 40

21. 36
22. 27
23. 72
24. 39
25. 36
26. 45
27. 34
28. 22
29. 16
30. 24
31. 28
32. 15
33. 12
34. 20
35. 16
36. 36
37. 40
38. 20
39. 27
40. 42
41. 12
42. 30
43. 21
44. 16
45. 10
46. 24
47. 12
48. 30
49. 14
50. 8
51. 48
52. 36

Lesson 36/pages 108-109

1. >
2. =
3. <
4. >
5. <
6. >
7. <
8. >
9. >
10. <
11. >
12. <
13. <
14. =
15. >
16. <
17. =
18. >
19. >
20. >
21. <
22. >
23. <
24. =
25. $\frac{3}{5}, \frac{2}{5}, \frac{1}{5}$
26. $\frac{6}{8}, \frac{5}{8}, \frac{3}{8}$
27. $\frac{11}{12}, \frac{10}{12}, \frac{7}{12}$
28. $\frac{1}{2}, \frac{1}{4}, \frac{1}{6}$
29. $\frac{7}{15}, \frac{2}{5}, \frac{2}{15}$
30. $\frac{6}{15}, \frac{1}{3}, \frac{1}{5}$
31. $\frac{4}{5}, \frac{11}{20}, \frac{2}{5}$
32. $\frac{4}{5}, \frac{11}{15}, \frac{2}{3}, \frac{1}{3}$
33. $\frac{13}{14}, \frac{5}{7}, \frac{4}{7}, \frac{1}{2}$
34. $\frac{5}{6}, \frac{5}{3}, \frac{2}{3}, \frac{1}{2}$
35. $\frac{1}{2}, \frac{3}{5}, \frac{2}{7}, \frac{1}{4}$
36. $\frac{23}{28}, \frac{3}{4}, \frac{5}{7}, \frac{2}{3}$
37. $\frac{3}{5}, \frac{17}{35}, \frac{3}{7}, \frac{2}{5}$
38. $\frac{9}{10}, \frac{1}{2}, \frac{2}{5}, \frac{1}{5}$
39. $\frac{6}{7}, \frac{5}{6}, \frac{5}{7}, \frac{21}{42}$

Lesson 37/pages 110-112

1. $\frac{6}{7}$
2. $\frac{1}{6}$
3. Troy
4. $\frac{3}{4}$
5. $\frac{2}{3}$
6. Car B
7. Screw Y
8. A
9. $\frac{2}{5}$
10. equal to
11. $\frac{1}{4}$
12. $\frac{7}{11}$
13. $\frac{7}{10}$
14. $\frac{8}{12} = \frac{2}{3}$
15. $\frac{36}{12}$ or $\frac{3}{1}$
16. $\frac{5}{8}$
17. $\frac{320}{8} = \frac{40}{1}$
18. $\frac{50}{10} = \frac{5}{1}$
19. $\frac{75}{100} = \frac{3}{4}$
20. $\frac{1}{3}$

Posttest Unit 5/page 113

1. $\frac{1}{5}$
2. $\frac{2}{8} = \frac{1}{4}$
3. $\frac{2}{4} = \frac{1}{2}$
4. $\frac{3}{3} = 1$

5. yes
6. no
7. yes
8. no
9. $\frac{4}{5}$
10. $\frac{5}{6}$
11. $\frac{1}{3}$
12. $\frac{1}{2}$
13. 12
14. 18
15. 30
16. 24
17. >
18. <
19. =
20. <
21. 6
22. 22
23. 33
24. $\frac{3}{12} = \frac{1}{4}$
25. $\frac{12}{16} = \frac{3}{4}$

Pretest Unit 6/page 114

1. $2\frac{1}{3}$
2. $4\frac{6}{8} = 4\frac{3}{4}$
3. >
4. =
5. <
6. =
7. $7\frac{5}{10} = 7\frac{1}{2}$
8. 8
9. $11\frac{1}{5}$
10. $4\frac{3}{12} = 4\frac{1}{4}$
11. $\frac{21}{4}$
12. $\frac{71}{5}$
13. $\frac{53}{8}$
14. $\frac{64}{15}$
15. $51\frac{1}{5}$
16. $241\frac{3}{9} = 241\frac{1}{3}$
17. $114\frac{33}{63} = 114\frac{11}{21}$
18. $32\frac{6}{24} = 32\frac{1}{4}$

Lesson 38/page 115

1. $1\frac{5}{8}$
2. $2\frac{4}{7}$
3. $2\frac{1}{6}$
4. $1\frac{5}{16}$
5. $2\frac{1}{2}$

Life Skill/pages 116-117
approximate answers

A. nail B 2 inches
B. bolt B $2\frac{1}{4}$ inches
C. wire B $2\frac{5}{16}$ inches
1. $3\frac{5}{8}$ inches
2. $2\frac{3}{8}$ inches
3. $4\frac{3}{4}$ inches
4. $1\frac{1}{8}$ inches
5. $2\frac{3}{4}$ inches
6. $4\frac{1}{2}$ inches
7. 5 inches
8. $1\frac{5}{8}$ inches

Lesson 39/pages 118-120

1.	=	19.	Nita
2.	>	20.	no
3.	<	21.	(3)
4.	<	22.	(2)
5.	<	23.	(4)
6.	>	24.	(4)
7.	>	25.	(4)
8.	=	26.	(3)
9.	>	27.	(4)
10.	=	28.	(4)
11.	<	29.	(2)
12.	=	30.	(3)
13.	>	31.	(3)
14.	=	32.	(4)
15.	<		
16.	>		
17.	>		
18.	>		

Lesson 40/pages 121-122

1.	$1\frac{5}{8}$	14.	$1\frac{1}{49}$
2.	$1\frac{9}{15} = 1\frac{3}{5}$	15.	$\frac{17}{5}$
3.	$2\frac{4}{8} = 2\frac{1}{2}$	16.	$\frac{33}{8}$
4.	$4\frac{4}{15}$	17.	$\frac{79}{9}$
5.	$3\frac{3}{6} = 3\frac{1}{2}$	18.	$\frac{58}{7}$
6.	$1\frac{17}{18}$	19.	$\frac{15}{15}$
7.	$2\frac{7}{9}$	20.	$\frac{20}{9}$
8.	$5\frac{2}{3}$	21.	$\frac{61}{8}$
9.	$2\frac{23}{49}$	22.	$\frac{31}{10}$
10.	$5\frac{4}{12} = 5\frac{1}{3}$	23.	$\frac{29}{6}$
11.	$3\frac{2}{3}$	24.	$\frac{126}{9}$
12.	3	25.	$\frac{47}{3}$
13.	$1\frac{4}{14} = 1\frac{2}{7}$	26.	$\frac{25}{8}$

Lesson 41/page 123

1.	$11\frac{1}{7}$	6.	$3\frac{3}{14}$
2.	$17\frac{1}{3}$	7.	$34\frac{4}{6} = 34\frac{2}{3}$
3.	$12\frac{1}{4}$	8.	$7\frac{2}{9}$
4.	$23\frac{6}{10} = 23\frac{3}{5}$	9.	$19\frac{4}{8} = 19\frac{1}{2}$
5.	$32\frac{4}{12} = 32\frac{1}{3}$		

Lesson 42/pages 124-125

1.	yes	5.	Louie
2.	more	6.	more
3.	Greg	7.	cars
4.	fiction	8.	liquid soap

Posttest Unit 6/page 126

1.	$2\frac{1}{2}$	5.	=
2.	$3\frac{2}{8} = 3\frac{1}{4}$	6.	>
3.	>	7.	$23\frac{1}{5}$
4.	<	8.	16

9. $19\frac{1}{2}$
10. 9
11. $\frac{32}{9}$
12. $\frac{45}{8}$
13. $\frac{39}{2}$
14. $\frac{63}{10}$
15. $297\frac{4}{8} = 297\frac{1}{2}$
16. $215\frac{2}{4} = 215\frac{1}{2}$
17. $100\frac{36}{72} = 100\frac{1}{2}$
18. $100\frac{30}{70} = 100\frac{3}{7}$
19. (3)
20. (2)

Pretest Unit 7/page 127

1.	$\frac{5}{24}$	9.	2
2.	$\frac{5}{16}$	10.	16
3.	$\frac{2}{3}$	11.	$\frac{8}{25}$
4.	$69\frac{3}{4}$	12.	$1\frac{2}{3}$
5.	12	13.	(4)
6.	4	14.	(3)
7.	$12\frac{1}{4}$	15.	(2)
8.	88	16.	(4)

Lesson 43/pages 128-129

1.	$\frac{1}{8}$	18.	$\frac{9}{20}$
2.	$\frac{1}{6}$	19.	$\frac{1}{30}$
3.	$\frac{1}{8}$	20.	$\frac{1}{12}$
4.	$\frac{1}{2}$	21.	$\frac{3}{20}$
5.	$\frac{1}{9}$	22.	$\frac{1}{15}$
6.	$\frac{3}{8}$	23.	$\frac{10}{27}$
7.	$\frac{1}{20}$	24.	$\frac{2}{35}$
8.	$\frac{2}{9}$	25.	$\frac{1}{32}$
9.	$\frac{3}{28}$	26.	$\frac{15}{32}$
10.	$\frac{7}{36}$	27.	$\frac{21}{50}$
11.	$\frac{1}{24}$	28.	$\frac{4}{15}$
12.	$\frac{3}{16}$	29.	$\frac{8}{15}$
13.	$\frac{1}{10}$	30.	$\frac{3}{50}$
14.	$\frac{1}{12}$	31.	$\frac{18}{35}$
15.	$\frac{35}{48}$	32.	$\frac{14}{27}$
16.	$\frac{3}{32}$	33.	$\frac{8}{33}$
17.	$\frac{8}{35}$		

Lesson 44/pages 130-131

1.	$\frac{1}{10}$	5.	$\frac{1}{9}$
2.	$\frac{3}{5}$	6.	$\frac{1}{7}$
3.	$\frac{3}{10}$	7.	$\frac{2}{3}$
4.	$\frac{1}{3}$	8.	$\frac{3}{5}$

9. $\frac{1}{8}$
10. $\frac{3}{4}$
11. $\frac{2}{9}$
12. $\frac{1}{8}$
13. $\frac{2}{7}$
14. $\frac{7}{10}$
15. $\frac{1}{4}$
16. $\frac{1}{3}$
17. $\frac{1}{8}$
18. $\frac{5}{16}$
19. $\frac{1}{3}$
20. $\frac{2}{25}$
21. $\frac{1}{3}$
22. $\frac{1}{6}$
23. $\frac{7}{30}$
24. $\frac{8}{15}$
25. $\frac{1}{12}$
26. $\frac{1}{8}$
27. $\frac{1}{4}$

Life Skill/page 132

1.	800 votes	4.	60 members
2.	4 million	5.	293 people
3.	33 states		

Lesson 45/pages 133-134

1.	28	21.	28
2.	6	22.	15
3.	12	23.	12
4.	5	24.	$13\frac{1}{2}$
5.	$\frac{2}{3}$	25.	13
6.	$\frac{5}{6}$	26.	$22\frac{1}{2}$
7.	$5\frac{1}{4}$	27.	12
8.	$1\frac{1}{3}$	28.	$7\frac{1}{2}$
9.	2	29.	24
10.	10	30.	35
11.	$17\frac{1}{2}$	31.	4
12.	10	32.	49
13.	$6\frac{2}{3}$	33.	$13\frac{1}{2}$
14.	50	34.	7
15.	$1\frac{4}{5}$	35.	28
16.	$4\frac{2}{7}$	36.	10
17.	$5\frac{7}{10}$	37.	250
18.	4	38.	47
19.	$\frac{6}{7}$	39.	$21\frac{1}{3}$
20.	6	40.	24
		41.	3
		42.	6

Life Skill/page 135

1. 4 ounces
2. $10\frac{2}{3}$ ounces
3. 9 doughnuts

Lesson 46/pages 136-138

1.	20	8.	44
2.	$6\frac{1}{3}$	9.	183
3.	$30\frac{1}{2}$	10.	$\frac{31}{64}$
4.	$1\frac{7}{20}$	11.	22
5.	$22\frac{1}{2}$	12.	675
6.	15	13.	$2\frac{1}{9}$
7.	$2\frac{1}{4}$	14.	$3\frac{1}{30}$

15. $6\frac{1}{4}$
16. $1\frac{1}{9}$
17. $5\frac{1}{4}$
18. 3
19. 215
20. $10\frac{1}{2}$
21. $15\frac{1}{6}$
22. 135
23. $28\frac{1}{2}$
24. 1
25. $3\frac{3}{20}$
26. 18
27. $14\frac{2}{3}$
28. 110
29. $\frac{17}{30}$
30. $8\frac{1}{6}$
31. $5\frac{3}{5}$
32. 33
33. 27
34. $1\frac{5}{6}$
35. 18
36. 9
37. 3
38. $14\frac{2}{7}$
39. $2\frac{3}{5}$
40. $\frac{1}{8}$ acre
41. 6 hours
42. 101 foreign imports
43. $12\frac{3}{8}$ miles
44. 715 miles
45. 450 ounces
46. $24.57
47. $105
48. 27 bags
49. $5\frac{5}{8}$ cups

Life Skill/page 139

Cobbler
36 cups pitted tart red cherries
$10\frac{1}{2}$ cups sugar
3 teaspoons grated lemon rind
$\frac{3}{4}$ teaspoon salt
3 pkg. piecrust mix
12 tablespoons butter

Yeast Rolls
1 cup milk
4 pkg. dry yeast
1 cup warm water
$1\frac{1}{2}$ cups butter or margarine
1 cup sugar
8 egg yolks
9 cups flour

Banana Bread
$3\frac{3}{4}$ pounds bananas
6 eggs
$\frac{3}{4}$ cup vegetable oil
$2\frac{1}{4}$ cups sugar
$2\frac{1}{4}$ cups shredded bran cereal
6 teaspoons baking powder
$1\frac{1}{2}$ teaspoons salt
$1\frac{1}{2}$ teaspoons baking soda
$3\frac{3}{4}$ cups flour
$1\frac{1}{2}$ cups shredded coconut
1 cup walnuts
$\frac{3}{8}$ cup pecans

Lesson 47/page 140

1. $1\frac{1}{5}$
2. $\frac{5}{6}$
3. 1
4. $1\frac{3}{4}$
5. 3
6. 2
7. $1\frac{1}{5}$
8. $\frac{5}{6}$
9. $\frac{4}{5}$
10. $\frac{2}{3}$
11. $\frac{2}{3}$
12. $1\frac{1}{4}$
13. 4
14. 3
15. $2\frac{2}{5}$
16. $5\frac{1}{3}$

Lesson 48/page 141

1. $\frac{1}{10}$
2. $\frac{1}{5}$
3. $\frac{1}{18}$
4. $\frac{3}{20}$
5. $\frac{1}{28}$
6. $\frac{1}{12}$
7. $\frac{5}{36}$
8. $\frac{2}{21}$
9. $\frac{5}{48}$
10. $\frac{1}{15}$
11. $\frac{1}{6}$
12. $\frac{1}{10}$
13. $\frac{1}{15}$
14. $\frac{1}{4}$
15. $\frac{1}{18}$
16. $\frac{1}{12}$

Lesson 49/page 142-143

1. $\frac{1}{9}$
2. $\frac{1}{10}$
3. 7
4. 7
5. $3\frac{1}{5}$
6. $4\frac{1}{6}$
7. $2\frac{1}{2}$
8. $6\frac{2}{3}$
9. $2\frac{1}{2}$
10. $\frac{2}{3}$
11. $1\frac{6}{11}$
12. $1\frac{3}{11}$
13. $\frac{5}{6}$
14. $1\frac{4}{5}$
15. 6
16. $4\frac{1}{5}$
17. 7
18. $3\frac{1}{3}$
19. $2\frac{2}{3}$
20. $1\frac{1}{2}$
21. $2\frac{4}{7}$
22. $\frac{1}{3}$
23. $4\frac{1}{12}$
24. 5
25. $3\frac{1}{5}$
26. $\frac{7}{15}$
27. 3
28. $1\frac{5}{16}$

29. >
30. >
31. <
32. >
33. <
34. <
35. =
36. <
37. <
38. <
39. <
40. <
41. <
42. <
43. =
44. >

Life Skill/page 144

1. $1\frac{7}{8}$
2. 1
3. 6
4. $\frac{9}{16}$
5. $\frac{9}{16}$
6. $\frac{1}{2}$
7. $\frac{1}{4}$
8. 1
9. $1\frac{1}{2}$
10. $\frac{1}{4}$
11. $1\frac{1}{4}$
12. $\frac{1}{3}$

Lesson 50/pages 145-147

1. $\frac{7}{8}$ yards
2. $\frac{3}{4}$ yards
3. $52\frac{6}{7}$ miles/hour
4. $2\frac{1}{4}$ hours
5. $\frac{1}{6}$ pie
6. $3\frac{3}{20}$ pounds
7. $8\frac{2}{3}$ miles/hour
8. 15 windows
9. $7\frac{4}{25}$ hours
10. 64 steaks

Life Skill/page 148

1. $437\frac{1}{2}$ miles
2. 2,125 miles
3. 1,125 miles
4. 375 miles

Posttest Unit 7/page 149

1. $\frac{1}{12}$
2. $\frac{2}{5}$
3. 28
4. 13
5. $14\frac{1}{4}$
6. $3\frac{3}{8}$
7. $\frac{2}{5}$
8. $\frac{3}{10}$
9. $\frac{1}{27}$
10. $1\frac{17}{60}$
11. $2\frac{5}{6}$
12. $5\frac{1}{3}$
13. $\frac{3}{4}$
14. $1\frac{1}{3}$
15. $\frac{1}{2}$
16. $1\frac{1}{20}$
17. $\frac{11}{50}$
18. $\frac{4}{21}$
19. 4
20. $1\frac{7}{9}$
21. $\frac{23}{32}$ pounds
22. 252 hours

Pretest Unit 8/page 150

1. $1\frac{2}{3}$
2. $\frac{1}{4}$
3. $\frac{33}{40}$
4. $6\frac{1}{3}$
5. $12\frac{13}{24}$
6. $1\frac{3}{5}$
7. $23\frac{1}{2}$
8. $3\frac{5}{12}$
9. $28\frac{1}{12}$ feet
10. $1\frac{1}{2}$ yards
11. $\frac{3}{4}$
12. $19\frac{5}{12}$ miles

Lesson 51/pages 151-152

1. $\frac{9}{11}$
2. 1
3. $1\frac{1}{4}$
4. $\frac{14}{15}$
5. $\frac{1}{8}$
6. $\frac{3}{5}$
7. $\frac{1}{5}$
8. $\frac{1}{2}$
9. $1\frac{1}{5}$
10. $1\frac{1}{5}$
11. $1\frac{2}{15}$
12. $\frac{7}{8}$
13. $\frac{2}{5}$
14. $\frac{1}{5}$
15. 0
16. $\frac{2}{3}$
17. $1\frac{1}{2}$
18. $\frac{9}{10}$
19. $1\frac{3}{11}$
20. $1\frac{3}{4}$
21. $\frac{1}{4}$
22. $\frac{1}{2}$
23. $\frac{1}{2}$
24. $\frac{3}{8}$
25. 1
26. $1\frac{1}{2}$
27. $\frac{4}{5}$
28. $\frac{2}{3}$
29. $\frac{2}{3}$
30. $\frac{2}{3}$
31. $\frac{1}{5}$
32. $\frac{1}{8}$

Lesson 52/pages 153-154

1. $\frac{5}{24}$
2. $\frac{17}{36}$
3. $1\frac{2}{15}$
4. $1\frac{4}{35}$
5. $\frac{1}{3}$
6. $\frac{1}{5}$
7. $\frac{1}{20}$
8. $\frac{3}{4}$
9. $\frac{23}{40}$
10. $1\frac{1}{12}$
11. $\frac{15}{16}$
12. $1\frac{4}{9}$
13. $\frac{2}{5}$
14. $\frac{5}{18}$
15. $\frac{2}{15}$
16. $\frac{27}{100}$
17. $\frac{19}{24}$
18. $1\frac{1}{24}$
19. $1\frac{1}{7}$
20. $1\frac{17}{30}$
21. $\frac{4}{21}$
22. $\frac{1}{24}$
23. $\frac{8}{15}$
24. $\frac{3}{10}$
25. $\frac{19}{20}$
26. $\frac{8}{9}$
27. $1\frac{13}{15}$
28. $1\frac{13}{24}$
29. $\frac{11}{24}$
30. $\frac{13}{18}$
31. 0
32. 0

Lesson 53/pages 155-156

1. $13\frac{1}{2}$
2. 10^2
3. $13\frac{7}{9}$
4. 43
5. $9\frac{2}{3}$
6. $6\frac{4}{7}$
7. $12\frac{1}{3}$
8. $20\frac{5}{16}$
9. 7
10. $16\frac{5}{7}$
11. 14
12. $6\frac{5}{6}$
13. $15\frac{4}{5}$
14. 13
15. 31
16. $10\frac{1}{3}$
17. $4\frac{1}{5}$
18. $15\frac{1}{3}$
19. $11\frac{1}{4}$
20. $16\frac{5}{12}$
21. $5\frac{9}{10}$
22. 22
23. $9\frac{1}{4}$
24. 22
25. $23\frac{2}{3}$
26. $21\frac{2}{5}$

Lesson 54/pages 157-158

1. $29\frac{9}{16}$
2. $10\frac{11}{30}$
3. $29\frac{1}{12}$
4. $19\frac{11}{12}$
5. $4\frac{19}{20}$
6. $113\frac{17}{24}$
7. $98\frac{8}{15}$
8. $11\frac{5}{9}$
9. $9\frac{11}{36}$
10. $18\frac{9}{20}$
11. $11\frac{9}{10}$
12. $11\frac{1}{24}$
13. $62\frac{9}{11}$
14. $112\frac{5}{6}$
15. $17\frac{16}{21}$
16. $25\frac{1}{16}$
17. $24\frac{3}{4}$ inches
18. $301\frac{3}{4}$ pounds
19. $26\frac{3}{4}$ pounds
20. $10\frac{3}{8}$ yards
21. $2\frac{5}{12}$ hours
22. $1\frac{1}{12}$ pounds

Life Skill/page 159

1. $11\frac{1}{2}$
2. $8\frac{1}{6}$ hours
3. $10\frac{1}{3}$ hours
4. $11\frac{1}{4}$ hours
5. 8 hours
6. $10\frac{3}{4}$ hours

Lesson 55/pages 160-161

1. $1\frac{3}{8}$
2. $2\frac{2}{3}$
3. $4\frac{2}{3}$
4. $2\frac{5}{6}$
5. $1\frac{2}{3}$

6. $16\frac{1}{3}$
7. $7\frac{11}{16}$
8. $3\frac{2}{5}$
9. $13\frac{3}{14}$
10. $8\frac{11}{18}$
11. $5\frac{13}{15}$
12. $27\frac{4}{9}$
13. $5\frac{2}{3}$
14. $2\frac{1}{3}$
15. $3\frac{1}{2}$
16. $15\frac{3}{10}$
17. $4\frac{1}{6}$
18. $3\frac{1}{3}$
19. $\frac{1}{2}$
20. $5\frac{3}{8}$
21. $9\frac{2}{5}$
22. $2\frac{5}{8}$
23. $7\frac{1}{6}$
24. $2\frac{5}{12}$
25. $\frac{7}{8}$
26. $2\frac{4}{7}$
27. $6\frac{7}{15}$
28. $2\frac{2}{3}$

Lesson 56/pages 162-163

1. $2\frac{7}{20}$
2. $7\frac{16}{21}$
3. $9\frac{1}{3}$
4. $4\frac{5}{16}$
5. $4\frac{2}{21}$
6. $4\frac{5}{16}$
7. $3\frac{3}{4}$
8. $9\frac{2}{15}$
9. $2\frac{9}{20}$
10. $4\frac{13}{16}$
11. $33\frac{7}{12}$
12. $2\frac{5}{8}$
13. $6\frac{5}{6}$
14. $3\frac{17}{25}$
15. $4\frac{11}{20}$
16. $4\frac{5}{12}$
17. $3\frac{19}{24}$
18. $6\frac{7}{16}$
19. $4\frac{1}{4}$
20. $1\frac{5}{12}$
21. $4\frac{1}{4}$
22. $2\frac{5}{6}$
23. $4\frac{11}{24}$
24. $4\frac{13}{15}$

Lesson 57/pages 164-166

1. $\frac{3}{8}$ inches
2. $3\frac{1}{2}$ yards
3. $2\frac{2}{3}$ gallons
4. $21\frac{1}{4}$ yards
5. $5\frac{3}{4}$ miles
6. $14\frac{3}{4}$ gallons
7. $\frac{5}{8}$ cup
8. $1\frac{1}{2}$ miles

Posttest Unit 8/pages 167-168

1. $\frac{3}{4}$
2. $\frac{3}{4}$
3. $2\frac{1}{12}$
4. $6\frac{3}{7}$
5. $4\frac{11}{12}$
6. $12\frac{1}{10}$
7. $6\frac{7}{10}$
8. $\frac{2}{3}$
9. $4\frac{1}{2}$
10. $6\frac{11}{16}$
11. $\frac{5}{12}$
12. $5\frac{1}{3}$

13. $1\frac{2}{3}$ quarts
14. $40\frac{1}{2}$ yards
15. $4\frac{1}{4}$ cups
16. $\frac{7}{12}$ pounds
17. (2)
18. (4)
19. (4)
20. (4)
21. (3)
22. (2)

Pretest Unit 9/page 169

1. 8.9
2. $4\frac{1}{10}$
3. 2.12
4. $5\frac{9}{100}$
5. .875
6. $\frac{5}{8}$
7. =
8. <
9. >
10. <
11. =
12. <
13. $11\frac{7}{8}$ or 11.875 yards
14. .375 or $\frac{3}{8}$ inches

Lesson 58/pages 170-171

1. $\frac{1}{10}$
2. $\frac{1}{100}$
3. $\frac{89}{1000}$
4. $\frac{8}{25}$
5. $\frac{1}{200}$
6. $\frac{1}{125}$
7. $1\frac{3}{10}$
8. $7\frac{1}{5}$
9. $\frac{8}{25}$
10. $\frac{1}{1000}$
11. $14\frac{3}{10}$
12. $310\frac{9}{10}$
13. $6\frac{147}{200}$
14. $29\frac{29}{100}$
15. $\frac{503}{1000}$
16. $65\frac{3}{50}$
17. $11\frac{1}{4}$
18. $20\frac{3}{4}$
19. $118\frac{1}{3}$
20. $\frac{9}{40}$

Lesson 59/pages 172-173

1. .1
2. .01
3. .2
4. .4
5. .5
6. .6
7. .7
8. .8
9. .9
10. .25
11. .75
12. $.33\frac{1}{3}$
13. $.66\frac{2}{3}$
14. .125
15. .375
16. .625
17. .875
18. $.16\frac{2}{3}$
19. $.83\frac{1}{3}$
20. .3
21. 2.2 hours
22. 8.5 hours
23. $5.66\frac{2}{3}$ hours
24. 3.8 hours

Lesson 60/page 174

1. .3
2. .8
3. .875
4. .0625
5. <
6. <
7. >
8. >

Lesson 61/page 175

1. <
2. >
3. <
4. =
5. >
6. =
7. <
8. <
9. <
10. <
11. <
12. =

Life Skill/page 176

1. $218.75
2. $237.50
3. $306.25
4. $327.25
5. $320.00
6. $410.00

Lesson 62/pages 177-178

1. 6
2. 18
3. 21
4. 9
5. 108
6. 45
7. 0
8. 6
9. 8
10. 36

Lesson 63/pages 179-180

1. 593.2 miles
2. 20 books
3. $\frac{1}{4}$
4. 1
5. 23.75 pounds

Posttest Unit 9/page 181

1. .3
2. $\frac{1}{5}$
3. $\frac{9}{20}$
4. .375
5. $\frac{3}{5}$
6. .04
7. $\frac{1}{125}$
8. .025
9. 1.5
10. 12.75
11. <
12. <
13. >
14. >
15. <
16. >
17. >
18. >
19. <
20. >
21. 7.625 or $7\frac{5}{8}$ pounds
22. 12.125 or $12\frac{1}{8}$ pounds

Practice for Mastering Fractions/pages 182-186

1. $\frac{1}{4}$
2. $\frac{15}{56}$
3. $\frac{63}{80}$
4. $10\frac{1}{2}$
5. $\frac{2}{3}$
6. 6
7. 136
8. 58
9. 27
10. $4\frac{8}{9}$
11. 69
12. $87\frac{1}{3}$
13. 11
14. $17\frac{1}{2}$
15. $19\frac{3}{8}$

16. $\frac{3}{4}$
17. $\frac{4}{5}$
18. $1\frac{1}{3}$
19. $2\frac{2}{5}$
20. $\frac{7}{45}$
21. $\frac{7}{100}$
22. 9
23. $2\frac{2}{3}$
24. $12\frac{4}{5}$
25. $1\frac{1}{2}$
26. $\frac{27}{32}$
27. $\frac{4}{7}$
28. $1\frac{4}{5}$
29. $1\frac{9}{14}$
30. $4\frac{1}{9}$
31. $9
32. 21 ounces
33. $11\frac{1}{4}$ pounds
34. 21 bags
35. 75 customers
36. $7\frac{1}{6}$ feet
37. $6\frac{1}{2}$
38. $8\frac{1}{12}$
39. $116\frac{23}{24}$
40. $4\frac{9}{10}$
41. $11\frac{31}{36}$
42. $14\frac{11}{16}$
43. $13\frac{31}{32}$
44. $16\frac{11}{18}$
45. $10\frac{3}{8}$
46. $22\frac{19}{24}$
47. $15\frac{3}{7}$
48. $51\frac{9}{16}$
49. $47\frac{5}{8}$ acres
50. $5\frac{3}{4}$ pounds
51. $\frac{13}{15}$
52. $1\frac{3}{16}$
53. $3\frac{1}{4}$
54. $3\frac{4}{9}$
55. $3\frac{17}{18}$
56. $2\frac{3}{4}$
57. $\frac{17}{24}$
58. $4\frac{3}{5}$
59. $3\frac{5}{9}$
60. $7\frac{1}{3}$
61. $\frac{22}{45}$
62. $43\frac{21}{40}$

63. $2\frac{1}{2}$ cups
64. $15\frac{17}{24}$ yards
65. (5)
66. (1)
67. (2)
68. (1)
69. (5)
70. (2)
71. (1)
72. (1)
73. (2)
74. (3)
75. (3)
76. (3)
77. (1)
78. (1)
79. (5)
80. (5)

Fractions Posttest/pages 187-190

1. $1\frac{1}{2}$
2. $\frac{1}{3}$
3. $\frac{3}{4}$
4. 3
5. $2\frac{1}{7}$
6. $\frac{23}{3}$
7. $7\frac{1}{2}$
8. 4
9. $.33\frac{1}{3}$
10. .2
11. 2.5
12. 3.375
13. $\frac{7}{8}$
14. $7\frac{1}{2}$
15. $\frac{3}{10}$
16. $\frac{2}{3}$
17. $=$
18. $<$
19. $=$
20. $=$
21. $<$
22. $=$
23. $<$
24. $>$
25. $14\frac{2}{7}$
26. $12\frac{1}{3}$
27. $28\frac{2}{3}$
28. $223\frac{3}{5}$
29. $.56; .625; .65; .8; 2.66\frac{2}{3}$
30. $\frac{1}{2}, \frac{1}{3}, \frac{1}{4}, \frac{1}{5}, \frac{1}{8}$

31. $\frac{1}{2}$
32. $3\frac{3}{4}$
33. $1\frac{3}{4}$
34. 20
35. 12
36. $7\frac{1}{5}$
37. $2\frac{1}{2}$
38. $\frac{5}{72}$
39. 20
40. $\frac{2}{5}$
41. $1\frac{4}{5}$
42. $1\frac{5}{6}$
43. $\frac{3}{10}$
44. 44
45. $1\frac{1}{5}$
46. $14\frac{19}{20}$
47. $13\frac{1}{12}$
48. $3\frac{23}{24}$
49. $13\frac{13}{24}$
50. $9\frac{37}{42}$
51. $58\frac{19}{20}$
52. $12\frac{23}{44}$
53. $35\frac{2}{3}$
54. $5\frac{1}{6}$
55. $7\frac{3}{14}$
56. $2\frac{23}{30}$
57. $15\frac{3}{10}$
58. $4\frac{1}{3}$
59. $2\frac{2}{3}$
60. $6\frac{1}{3}$
61. $=$
62. $<$
63. $>$
64. $>$
65. (1) $\frac{20}{25} = \frac{3}{5}$
66. (3) $2\frac{1}{2}$
67. (3) $13\frac{5}{8}$ yards
68. (2) 11.625 inches
69. (3) $\frac{1}{5}$
70. (3)
71. (2) 5 aprons
72. (2) $1\frac{1}{12}$ feet
73. (4) $6\frac{1}{2}$ hours
74. (3) $6\frac{1}{2}$ miles
75. (2) $3\frac{1}{4}$ pounds
76. (3) 4.25 hours
77. (3) 15.125 inches
78. (1) $\frac{2}{3}$

Check 1

Step 1 → _April 14_ 19 _95_ 7-15/520

Step 2

Step 3

PAY TO THE ORDER OF _Mutual Life Insurance_ $ _26.13_

Twenty-six and 13/100 ——— DOLLARS

City National Bank
Annapolis

Step 4

Memo _Policy #632_   _Betty Adair_

⑆052000159⑆ 6343019⑈   Step 5

Check 2

_____ 19 ___ 7-15/520

PAY TO THE ORDER OF _Forest Hill Apartment_ $ _353.70_

Three hundred fifty-three and 70/100 DOLLARS

City National Bank
Annapolis

Memo _rent_

⑆052000159⑆ 6343019⑈

Check 3

_____ 19 ___ 7-15/520

PAY TO THE ORDER OF _Commonwealth Edison_ $ _56.39_

Fifty-six and 39/100 ——— DOLLARS

City National Bank
Annapolis

Memo _electric bill_

⑆052000159⑆ 6343019⑈

Check 4

_____ 19 ___ 7-15/520

PAY TO THE ORDER OF _National Telephone Co._ $ _44.17_

Forty-four and 17/100 ——— DOLLARS

City National Bank
Annapolis

Memo _phone bill_

⑆052000159⑆ 6343019⑈

CASH	CURRENCY	25	00
	COIN	9	60
LIST CHECKS SINGLY		15	03
		50	00
		143	73
TOTAL FROM OTHER SIDE		—	—
TOTAL		243	36
▶ LESS CASH RECEIVED		130	00
NET DEPOSIT		113	36

CASH	CURRENCY	157	00
	COIN	3	89
LIST CHECKS SINGLY		59	65
		28	09
TOTAL FROM OTHER SIDE		—	—
TOTAL		248	63
▶ LESS CASH RECEIVED		200	00
NET DEPOSIT		48	63

GUEST CHECK

TABLE NO.	NO. PERSONS	CHECK NO.	SERVER NO.
		80466	

1	Lamb Chops	7	25
2	Child's Plate	7	00
1	Duckling	4	25
1	Sirloin Steak	7	25
1	Fried Chicken	5	75
2	Vegetable Soup	1	90
2	Coffee	1	20
2	Milk, small	1	60
2	Soft drinks, large	2	00
2	Apple Pies	2	50
2	Chocolate Cakes	2	50
2	Ice Cream	2	50
	Subtotal	45	70
TAX			

CHARLES KELLER

TONGUE TWISTERS

ILLUSTRATED BY RON FRITZ

Simon and Schuster Books for Young Readers
Published by Simon & Schuster Inc., New York

SIMON AND SCHUSTER BOOKS FOR YOUNG READERS
Simon & Schuster Building
Rockefeller Center
1230 Avenue of the Americas
New York, New York 10020

SIMON AND SCHUSTER BOOKS FOR YOUNG READERS
is a trademark of Simon & Schuster Inc.

Manufactured in the United States

10 9 8 7 6 5 4 3 2 1
10 9 8 7 6 5 4 3 2 1 pbk.

Library of Congress Cataloging-in-Publication Data
Keller, Charles.
Tongue twisters.
Summary: An illustrated collection of tongue twisters and other hard-to-say rhymes.
1. Tongue twisters. [1. Tongue twisters] I. Fritz, Ronald, ill. II. Title.
PN6371.5.K44 1989   818'.5402 88-26448

ISBN 0-671-67123-5
ISBN 0-671-67975-9 pbk.

To Nicole and Leigh
C. K.

To my wife Martha
R. F.

Betty bought some butter, "But," she said, "this butter's bitter, and a bit of better butter would make a better batter." So she bought a bit of butter better than the bitter butter, and it made her batter better — so it was that Betty bought a bit of better butter!

The sixth sheik's sixth sheep's sick.

Sheep shouldn't sleep in a shack.
Sheep should sleep in a shed.

Seven silly sheep slowly shuffled south.

A noise annoys an oyster, but a noisy noise annoys an oyster more.

S piral-shelled sea snails shuffle in sea shells.

E ight apes ate eight apples.

T hree tree toads tied together tried to trot to town.

Eight great gray geese grazing gaily in Greece.

A big black bug bit
a big black bear,
making the big black bear
bleed blood.

A haddock, a haddock,
a black spotted haddock.
A black spot on the
black back of
a black spotted haddock.

Peter Piper picked a peck of pickled peppers;
A peck of pickled peppers Peter Piper picked.
If Peter Piper picked a peck of pickled peppers,
Where's the peck of pickled peppers Peter Piper picked?

She sells sea shells on the seashore.
The shells she sells are sea shells, I'm sure.
And if she sells sea shells on the seashore,
Then I'm sure she sells seashore shells.

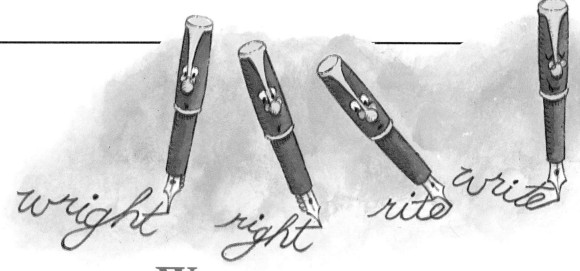

W rite, we know, if written right,
Should not be written wright or right,
Nor should it be written rite, but write,
For only then is it written right.

W hether the weather be fine,
Or whether the weather be not,
Whether the weather be cold,
Or whether the weather be hot,
We'll weather the weather,
whatever the weather,
Whether we like it or not.

Amidst the mists and coldest frosts,
With barest wrists and stoutest boasts,
He thrusts his fists against the posts
And still insists he sees the ghosts.

Lucy loosened Suzie's shoes and Suzie's shoes
stayed loose while Suzie snoozed.

Isabella broke the black umbrella.

Sister Sarah shined her silver shoes for Sunday.

Sheila Shorter sought a suitor;
Sheila sought a suitor short.
Sheila's suitor's sure to suit her;
Short's the suitor Sheila sought!

Some seventy-six sad, seasick seamen soon set sail, seeking soothing, salty South Sea sunshine.

Shallow ships show some signs of sinking.

Seven shy sailors salted salmon shoulder to shoulder.

You've no need to light a night light
On a light night like tonight,
For a night light's light's a slight light,
And tonight's a night that's light.
When a night's light, like tonight's light,
It's really not quite right
To light night lights with their slight lights,
On a light night like tonight.

Six thick thistle sticks.

Fluffy finches flying fast.

Nimble noblemen nibbling nuts.

Ellen's elegant elephant.

This is a zither.

Terrified tomcats
in the tops of
tall trees.

A maid with a duster made a furious bluster
Dusting a bust in the hall.
When the bust, it was dusted,
The bust, it was busted.
The bust, it was dust —
That is all!

A tutor who tooted a flute,
Tried to teach two tooters to toot.
Said the two to the tutor,
"Is it harder to toot,
Or tutor two tooters to toot?"

Old oily Ollie oils old oily autos.

Martin met a mob of marching munching monkeys.

The bootblack brought the black book back.

Frank threw Fred three free throws.

Pop keeps a lollipop shop
and the lollipop shop keeps pop.

Sheila uttered a sharp shrill shriek
and shrunk from the shriveled form
that slumbered in the shadows.

Freddie's friend Eddie phoned
for Freddie to fetch fruit from the farm
of the famous French farmer.

Hard-hearted Harold hit Henry hard
with a hickory-handled iron hammer.
Henry howled horribly and hurriedly
hobbled home.

Round and round the rough
and ragged rock the ragged rascal ran.

How much wood could a woodchuck chuck,
If a woodchuck could chuck wood?
It would chuck as much wood as woodchuck could,
If a woodchuck could chuck wood.

A fly and a flea in a flue,
were imprisoned,
So what could they do?
Said the flea, "Let us fly."
Said the fly, "Let us flee."
So they flew through a
flaw in the flue.

"Go, my son, and shut the shutter."
This I heard a mother utter.
"Shutter's shut," the boy did mutter.
"I can't shut it any shutter."